CONSTRUCTION DISPUTE RESOLUTION HANDBOOK

Robert Silver

Gary T. Furlong

Construction Dispute Resolution Handbook
© LexisNexis Canada Inc. 2004
October 2004

All rights reserved. No part of this publication may be reproduced, stored in any material form (including photocopying or storing it in any medium by electronic means and whether or not transiently or incidentally to some other use of this publication) without the written permission of the copyright holder except in accordance with the provisions of the Copyright Act. Applications for the copyright holder's written permission to reproduce any part of this publication should be addressed to the publisher.

Warning: The doing of an unauthorized act in relation to a copyrighted work may result in both a civil claim for damages and criminal prosecution.

Members of the LexisNexis Group worldwide

Canada	LexisNexis Canada Inc, 123 Commerce Valley Drive, MARKHAM, Ontario
Argentina	Abeledo Perrot, Jurisprudencia Argentina and Depalma, BUENOS AIRES
Australia	Butterworths, a Division of Reed International Books Australia Pty Ltd, CHATSWOOD, New South Wales
Austria	ARD Betriebsdienst and Verlag Orac, VIENNA
Chile	Publitecsa and Conosur Ltda, SANTIAGO DE CHILE
Czech Republic	Orac sro, PRAGUE
France	Éditions du Juris-Classeur SA, PARIS
Hong Kong	Butterworths Asia (Hong Kong), HONG KONG
Hungary	Hvg Orac, BUDAPEST
India	Butterworths India, NEW DELHI
Ireland	Butterworths (Ireland) Ltd, DUBLIN
Italy	Giuffré, MILAN
Malaysia	Malayan Law Journal Sdn Bhd, KUALA LUMPUR
New Zealand	Butterworths of New Zealand, WELLINGTON
Poland	Wydawnictwa Prawnicze PWN, WARSAW
Singapore	Butterworths Asia, SINGAPORE
South Africa	Butterworth Publishers (Pty) Ltd, DURBAN
Switzerland	Stämpfli Verlag AG, BERNE
United Kingdom	Butterworths Tolley, a Division of Reed Elsevier (UK), LONDON, WC2A
USA	LexisNexis, DAYTON, Ohio

Library and Archives Canada Cataloguing in Publication

Silver, Robert

Construction dispute reolution handbook / Robert Silver, Gary T. Furlong.

ISBN 0-433-44478-9

1. Construction contracts — Canada. 2. Dispute resolution (Law) — Canada.
I. Furlong, Gary T. II. Title.

KE933.S58 2004   343.71'078624   C2004-905664-6
KF902.S58 2004

Printed and bound in Canada.

Foreword

The famous American comedian W.C. Fields once said that he would never go into a building with a blindfolded woman standing outside with a set of scales in her hand. The statue of the woman was, of course, the symbol of Court Justice.

If Mr. Fields had read this book, he might very well have found no need to enter the building he described.

The book covers more than just dispute resolution; it deals also with prevention of disputes, which is so vitally important in the construction industry.

It is written in the language of the construction industry, and gives very valuable insight into the psychological mosaic of the participants in the process.

In my view, this book should be placed beside the plans and specification in every job shack, on every construction site, and be considered mandatory reading.

My congratulations to the authors on a very unique book, which I am sure will be enthusiastically received by all those who are involved in construction.

David I. Bristow, Q.C.

Toronto, August 2004

Preface

The construction industry has an important place in human history dating back to the origins of man, creating some of the greatest Wonders of the World in projects such as the Great Pyramids and the Taj Mahal. In modern times, construction has evolved into one of the most important industries in our national economy, employing thousands and literally building our cities, countries, and societies.

Historically, disputes that arose in the construction industry (as in many areas) were resolved in some fashion directly between the parties, either amicably on a handshake over a drink, or less than amicably through threats or violence.

There has been a shift in the last 20 to 30 years to litigation as the default process for resolving conflict in the construction industry. Indeed, as a society we have become more and more litigious, relying on the courts to enforce our rights as the favoured way to resolve a dispute.

Litigation, however, has not met the needs of the construction industry. In an industry where time is not only money but a great deal of money, the time it takes to either sue the other side and win or to resolve lien claims through the courts makes litigation far too expensive. In addition, the litigation process itself actually discourages the parties from resolving disputes on their own. By restricting and channelling all communications through counsel, and by using examinations for discovery and formal documentary exchange to learn the basic facts of the situation, virtually all contact between the parties is eliminated. Indeed, the first time that all parties are in the same room together after a claim is filed is at the trial itself, so it should be no surprise that most voluntary settlements occur on the courthouse steps, sometimes two, three, even four or more years after the issues arise.

Given these issues and concerns with litigation, a new approach and set of processes were developed that are now referred to as "Alternative Dispute Resolution", or ADR. ADR refers to a wide range of processes starting with negotiation and moving through mediation to arbitration with many different variations in between. The effective use of ADR (and mediation in particular) brings forward the promise of faster, cheaper resolutions of disputes along with better, more creative solutions; in addition, these processes offer the chance of maintaining, even improving the relationship between the parties, something of great value in the construction industry.

It should be noted that the use of ADR and mediation is not a panacea or magic bullet that solves all problems. While mediating a construction dispute may not always provide a final resolution for all of the issues involved, it can help the parties identify and meet their most important needs. If the construction project is still underway, an effective mediation process may allow for the completion of the project, keeping the delay claims and expenses associated with these claims to a minimum. If the project is already completed, a well-run arbitration process can bring closure much faster, as well as deliver a decision that the parties see as fairer, given that the arbitrator(s) is often accepted as a credible expert in the field.

When we started looking into writing this book, we discovered that while ADR is used frequently in many different industries, there are very few books that look at best practices for ADR specifically in the construction field. While there are many books on construction law and the lien act, there were very few that spoke to how construction professionals could effectively learn and implement the full range of ADR options, appropriate to the situation. After speaking with contractors, lawyers, court masters, building departments, financial institutions, trades and sub trades over the years we knew that this field would welcome new ways to deal with their issues that avoided litigation and helped to maintain business relationships. ADR offers useful and effective approaches for doing this.

Our objective in writing this book is to provide a useful understanding of how to deal effectively with disputes using a variety of ADR tools, and to provide a full range of best practices to accomplish this in the construction industry in three phases — before breaking ground, while building the project, and after a project is completed.

This book is designed to give lawyers, engineers, architects, designers, builders, customers, contractors, subcontractors, suppliers, trades, building departments, financial institutions and anyone else who may practise or have an interest in the construction industry not only a sense of the types of issues that will arise and cause conflict on a construction project, but effective ways of either preventing that conflict or dealing with it when it arises.

Finally, the book is written to give all the parties involved in construction disputes a firm understanding of each other's perspective and how ADR and mediation can be used by lawyers and their clients to accomplish their objectives, protect their interests, and keep the project moving while preserving the relationships in the field.

For the sake of simplicity, we have designed this book to follow the flow of a building project. Chapters 1 and 2 look at the foundation, which include the dynamics of conflict and the principles of ADR; Chapters 3, 4, and 5 build on this by describing ADR best practices before, during and after the project (in that order); and Chapter 6 focuses specifically on best practices for mediating a construction dispute. Chapter 7 provides the finishing touches with a summary and a look to the future of ADR in the field.

In addition, we have included a number of case stories at the end of the chapters 3, 4, and 5 to put some of the best practices in the context of actual cases we have worked on. At the back of the book we have included a few appendices we thought might be of interest as a reference for the reader, including CCDC and CCA dispute resolution clauses, a sample Agreement to Mediate, a copy of a Partnering Charter developed during a Partnering workshop, and an extended bibliography.

This book is written from our combined 27 years of experience and knowledge in the construction industry, as well as the field of ADR as applied in the construction industry and beyond. We urge you to take the information provided in this book to help you formulate your own ways in which ADR and mediation can be of help to you or your clients.

ACKNOWLEDGMENTS

The authors would like to acknowledge and thank the following people who contributed and assisted with this book: David Bristow Q.C., Ken Selby, Master David Sandler, Superior Court of Justice, Stanley Naftolin, J.D., Q.C., Rick Russell, Gian-Luca Di Rocco and the staff at LexisNexis Canada. Your assistance was most appreciated.

Gary Furlong

Robert Silver

* * *

This book has been completed with the encouragement and assistance of many people. I would like to thank my spouse, Jane, and my children, Joel, Lee, Kyle and Emily for being understanding, while I was undertaking this project, which took away from some of our Family time together. I would also like to thank my sister Shelly Lee for her encouragement.

My heartfelt thanks are extended to those who participated directly in helping me bring this book to fruition. I am indebted to my co-author, Gary Furlong, for his insight and knowledge of alternative dispute resolution and relating it into the unique nature of construction disputes.

Robert Silver

I would especially like to thank my spouse, Ronalda, and my children, Callan and Tess, for their patience and support throughout the writing of this book, not only for the time away I spent writing, but also for the long hours spent listening to me discuss the many fascinating issues and cases the book covers.

Gary T. Furlong

June, 2004
Toronto, Canada

ABOUT THE AUTHORS

Robert Silver

Robert Silver is a builder and general contractor who began building a strong reputation in the custom building industry in 1988. He has been involved in high-end custom homes, residential development, renovations and project management. He has been responsible for construction management of numerous projects valued at more than $1M. His work has been featured on the television shows "HGTV's Let's Build" and the "Discovery Channel's Harrowsmiths Country."

Over the years he has provided conciliation services in the building industry, including mediating disputes between trades, subtrades, manufactures, homeowners, building departments and financial institutions.

Robert has taught courses on communication, negotiation, transactional analysis, interpersonal skills and account management. His also teaches courses on "How to Build" from the ground up and how to keep the "Conflict Poison" from flowing on a construction project.

Robert studied at George Brown College in the field of electronics and at York University in the field of Alternative Dispute Resolution. He is a member of the Alternative Dispute Resolution Institute of Ontario and the Alternative Dispute Resolution Section of the Ontario Bar Association.

Robert Silver was born and raised in Toronto, and now lives in Unionville, Ontario with his wife Jane. They have four children, Joel, Lee, Kyle and Emily.

Gary T. Furlong

Gary Furlong has been a leading mediator and conflict resolution specialist for more than a decade, and has extensive experience in mediation, negotiation, alternative dispute resolution, Partnering, and conflict resolution. Gary is past President of the ADR Institute of Ontario, and holds a Master of Laws (LL.M.) in ADR from Osgoode Hall Law School, as well as the Chartered Mediator designation from the ADR Institute of Canada. Gary is a fellow of the International Academy of Mediators, and has been a roster mediator for the Ontario Mandatory Mediation Program and Law Society of Upper Canada dispute resolution panel for a number of years.

In the construction field, Gary Furlong and Rick Russell co-facilitated some of the first construction Partnering workshops in Canada for clients such as the Department of National Defence, Defence Construction Canada, and the Ontario Ministry of Transportation. During the last decade, Gary has lead many Partnering workshops in the construction industry for major Canadian contractors, both design-build and design-bid-build. Gary has mediated construction disputes ranging from straight damage claims to construction lien claims.

Gary was born in Hamilton, and attended Stanford University in California. Gary lives in Toronto with his wife, Ronalda Jones, who is a writer in the film and television industry. They have two children, Callan and Tess.

TABLE OF CONTENTS

Foreword iii
Preface v
Acknowledgments ix
About the Authors xi

Chapter 1: Introduction
I. What is Conflict Poison in the Construction Industry? 3
II. Conflict Poison — How It Starts and Grows 4
III. Disputes in the Construction Industry 6
    (a) Different Types of Disputes 7
        (i) Timing 7
        (ii) Payment Schedule 8
        (iii) Substandard Quality 9
        (iv) Change Orders 9
    (b) The Structure of a Construction Project 9
        (i) Blue Prints 10
        (ii) Communication and Co-ordination throughout the Project 10
        (iii) Finishing and Closing a Project — The Construction Lien Act 11
    (c) Interest-based ADR 13
IV. Summary 14

Chapter 2: Dispute Resolution — An Overview
I. Interest-based Processes 15
    (a) Example of Interest-based Approach 15
II. Rights-based Processes 16
    (a) Example of Rights-based Approach 16
III. Power-based Processes 17
    (a) Example of Power-based Approach 17
IV. Interests, Rights and Power 18
    (a) Conflict Poison and Interests, Rights and Power 19
        (i) Barriers to Resolution: Why it is Difficult to Reverse the Flow of Poison 19
    (b) Dispute Resolution Processes: This ain't the Stairway to Heaven! 21
        (i) Prevention 22
        (ii) Negotiation 23
        (iii) Non-Binding Processes 23
        (iv) Standing Neutral 23
        (v) Binding Processes 24
        (vi) Litigation 24
    (c) Interest-based Processes Reverse the Flow of Conflict Poison .. 24

V. Summary 26

Chapter 3: Best ADR Practices — Prevention and Early Dispute Resolution

I. Prevention — The Best Medicine 27
   (a) Align Expectations — Be Comfortable with Who you Hire 27
   (b) Detailed Discussions about the Quotes 28
   (c) Check out References..29
   (d) Partnering 30
      (i) What is Partnering...31
      (ii) The Partnering Workshop 32
   (e) Issue-Resolution Process..33
   (f) Communicating vs. Decision Making......................................34
   (g) Key Contract Provisions 35
      (i) Who is the Contract Between?................................35
      (ii) What is the Address of the Project? 36
      (iii) Dispute Resolution Clauses.....................................36
      (iv) Standing Neutral..39
      (v) Risk Allocation..40
      (vi) Other Key Contract Provisions 41
          (A) Bid and Performance Bonds.........................41
          (B) Change Orders..41
          (C) Payment Stream and Invoicing.....................41
          (D) Project Insurance 41
          (E) Communication 42
          (F) Warranties 42
          (G) Deficiencies...42
          (H) Commencement and Completion Dates 42
          (I) Delays or Disclaimer Clauses........................42
          (J) Value Engineering..42
II. Summary 43
III. Case Studies 43
   (a) Case #1: Bridge Over Untroubled Waters................................43
   (b) Case #2: The Case of the Missing Contractor..........................45
      (i) Lessons from the Case..47

Chapter 4: Best ADR Practices — Dispute Resolution During the Construction Project

I. Communication 49
   (a) Daily Site Meetings...50
   (b) Weekly Site Meetings 52
      (i) Attendance..52
      (ii) Meeting Agendas 53
      (iii) Weekly Site Meeting Minutes..................................53
      (iv) Senior Management Meetings.................................57
      (v) Summary — Communication..................................58

II. Issue Resolution Process 59
   (a) Issue Resolution Ladder (IRL)...60
   (b) Using a Standing Neutral 64
   (c) Other Dispute Resolution Issues During the Project................64
      (i) Change Orders or Extras 64
      (ii) Value Engineering Process...65
III. Summary 66
IV. Case Studies 66
   (a) Case #1: Smokin' the Big Pipe 66
      (i) Lessons from the Case..68
   (b) Case #2: Now You See Me, Now You Don't! 68
      (i) Lessons from the Case..69

Chapter 5: Best ADR Practices — Dealing with Disputes at the End of the Project

I. Mediation 72
   (a) The Mediation Process..72
      (i) Step 1: Pre-mediation 73
      (ii) Step 2: Introduction and Opening Statements.....................74
      (iii) Step 3: Identifying the Issues 75
      (iv) Step 4: Exploring the Issues and Problem Solving 75
      (v) Step 5: Reaching and Documenting an Agreement............76
   (b) Substantive Knowledge of the Mediator...................................76
   (c) Pros and Cons of Mediation..77
II. Neutral Evaluation 79
   (a) Neutral Expert Opinion...79
   (b) Non-Binding Arbitration...80
   (c) Mini-trial 81
III. Arbitration...81
   (a) Pitfalls of Arbitration 82
      (i) Costs and Time..82
      (ii) Fairness and Predictability of Outcome 83
      (iii) Lack of Right to Appeal 85
IV. Best Practices in Arbitration 85
   (a) Strong Initial Focus on the Procedural Side of the
       Arbitration..86
   (b) Strong Focus on the Handling of Evidence to Minimize
       Time and Delays 86
   (c) Tight Control of the Hearing by the Arbitrator 87
V. Hybrid Processes 87
   (a) Med-Arb...87
   (b) Arb-Med..88
VI. Litigation...89
   (a) Costs..89
   (b) Complexity..90
VII. Summary 91

VIII. Case Studies 91
    (a) Case #1: If you Can't Stand the Heat, Get Out of the Kitchen .. 91
        (i) Lessons from the Case 93
    (b) Case #2: Pulling the Plug 93
        (i) Lessons from the Case 95

Chapter 6: The Seven Habits for Successfully Mediating Construction Disputes

I. Habit #1: Mediate Early 97
    (a) Mediating Late 97
    (b) Mediating Early 98
II. Habit #2: Prepare, Prepare, Prepare 98
    (a) Briefs 98
    (b) Client Preparation 99
    (c) Who Attends 100
III. Habit #3: See This as a Business Decision, Not a Legal Problem 101
IV. Habit #4: Use Visual Aids 101
V. Habit #5: Use the Mediator Effectively 102
VI. Habit #6: Think Outside the Box 103
VII. Habit #7: Document the Settlement Well 103
VIII. Summary 104

Chapter 7: The Future of Construction ADR

I. The Future of ADR in Construction Disputes 106
    (a) Government Intervention 106
    (b) Partnering and Other Preventative Approaches 108
        (i) Partnering 108
        (ii) Public/Private Partnerships 109
        (iii) Training and Skills 109
II. Summary 109

Appendix 1: Bibliography 111

Appendix 2: Draft Agreement to Mediate 115

Appendix 3: Town of New Bedford/Ontario Water Support Agency and Oak Engineering Sewage Infrastructure Upgrades 119

Appendix 4: Contractual Dispute Resolution Clauses 121

Index 127

1

INTRODUCTION

Construction is a complex, demanding and difficult industry. To complete a construction project effectively, large or small, requires planning, co-ordination, and teamwork from all concerned, and in many cases, even that is not enough. The construction process must also account for a large number of variables that are outside of the parties' knowledge or control, and which test everyone's ability to creatively adapt to new circumstances. Combine that with a large number of parties, tight timeframes and thin profit margins, and you can see why dispute resolution is an important part of the construction process.

To be successful in the construction industry, like it or not, you need to be a dispute resolution specialist. When working in construction, conflict is not a question of "if", but a question of "when". One of my friends sent this story, which shows that even the simple act of building a boat in your own backyard can be difficult, causing pressure and conflict for all parties involved.

The Process of Building

The Lord came to Noah, in Canada, in the year 2004. The earth was wicked and over-populated. The Lord instructed Noah to build another ark and save two of every living thing, along with a few good humans.

"Here's the blueprint, build it exactly to the plans," said the Lord. "And hurry — in six months I start the rain for 40 days and 40 nights."

Six months later the rain started. The Lord looked down and saw Noah weeping in his flooded yard, and no ark.

"Noah," He roared, "Where is the ark?"

"Forgive me, Lord," begged Noah. "Things have not been easy. First, I needed a building permit. I've been arguing with the inspector about the need for a sprinkler system. Then, my neighbors claimed that I have violated the neighbourhood zoning laws by building the ark in my yard and exceeding the height limitation. We had to go to the Development Appeal Board for a decision."

"Then Transport Canada, the Department of Highways, and Hydro wanted a bond posted for the future costs of moving power, trolley and other overhead obstructions, to clear the passage for the ark's move to the sea. I argued the sea would be coming to us, but they would hear nothing of this."

"Getting the wood was another problem. There's a ban on cutting local trees in order to save the spotted owl. I tried to convince the environmentalists that I needed the wood to *save* the owls. No go! When I finally got the wood and began building, a soils engineer told me that I needed to pour a concrete pad, because the clay soil would allow the ark to sink into the mud before it was finished."

"I gathered the animals, but then I got sued by an animal rights group. They insisted that I was confining wild animals against their will."

"Environment Canada decided that I could not build the ark without filing an environmental impact statement on your proposed flood. I'm still trying to resolve a complaint with the Human Rights Commission on how many minorities I'm supposed to hire for my building crew. And the trade unions want me to hire only union members with ark-building experience."

"To make matters worse, the bank refused to advance any more funds when they found out I might be leaving the country in the ark before the 45-day window for construction liens to be filed expires."

"So, forgive me, Lord, but it would take at least 10 years to finish this ark."

Suddenly the skies cleared and the sun began to shine. A rainbow stretched across the sky. Noah looked up in wonder.

"You mean you're not going to destroy the world?" he asked.

"No," said the Lord. "You humans beat me to it!"

Although exaggerated, this story is not as far-fetched as it seems given the type of unexpected problems and the many approvals needed in order to have a building project even break ground. The approval stage for permits is just the beginning; the potential for conflict in any construction project is almost unlimited.

In the construction industry delays are very costly. Disputes that are not resolved as early as possible are extremely detrimental to all parties involved. Alternative Dispute Resolution (ADR) methods can expedite the resolution of the disputes that invariably occur in the construction industry.

Every year almost 97 per cent of all claims issued in Ontario courts settle before trial. So why do so many construction disputes end up reaching the court house steps before seriously considering settlement? Part of the answer lies in the construction industry culture. It is a tough business where people who "come out swinging" are admired. The courts support these values, offering a place where you can conduct a form of virtual warfare — except that your weapons are words.

When a conflict develops, our first objective should be to discuss our interests, but because emotions are often running high we tend to forget that in any conflict each side has its own perspective. In any construction project, the owner, municipality, contractor, subcontractors, inspectors, financial institutions, architect, and the other trades involved all know that they will encounter problems along the way. Understanding all of the parties' issues and perspectives can help resolve disputes, ensuring that they do not impede the construction project at hand.

Understanding why these disputes arise and knowing how to resolve them can save you time, money and emotional stress. Building projects involve hundred of components each dependent upon such variables as manufacturers, suppliers, subtrades, supervisors, inspectors, strikes and weather, to name just a few. Excellent communication, co-ordination and timing are crucial for things to stay on schedule. When any of these components go out of synch, *and they will*, the "conflict poison" begins to flow.

I. WHAT IS CONFLICT POISON IN THE CONSTRUCTION INDUSTRY?

Every construction project has its own identity; even more, each project takes on a life of its own. It has a "heartbeat", and behaves like a living breathing entity that can create difficult issues on a moment's notice. "Conflict poison" is the amount of conflict (*i.e.*, negative energy) that exists within a given construction project. It is more powerful than all of the resources needed during the construction itself. Finding a way to control this conflict poison as it arises is imperative, and will determine whether you have a positive or negative outcome on the project.

The construction industry is a unique field, and the specific way in which disputes escalate to full blown problems can be compared to the flow of blood, which runs through our veins. If a small amount of poison or toxin is introduced into our body and detected early, it can usually be treated with medication in a way that allows us to go about our daily routines. If not dealt with early, if the poison is not stopped and treated, the poison can flow out of control and cause major problems.

On a construction project, for many of the reasons explained in this book, the project may be injected with conflict poison many times. Being able to detect this conflict poison on time and treat it is imperative; it prevents the slow buildup of negative energy and damaged relationships from flowing out of control. When the conflict escalates out of control, it can result in a complete breakdown on site and a stoppage of work on the project.

The general contractor and the customer usually form the "heart" of the project. These parties are very powerful and have the ability to either administer medication that controls the conflict poison, or to increase the flow of poison until it is out of control. In order for a construction project to stay on track, it is imperative that these parties continue to monitor the blood and the temperature of the project, ensuring the conflict poison is kept to a minimum. They can instill a positive or negative balance within a construction project.

While there is a vested interest for the parties at the heart of the project to have it go well, there are also specific situations where they see each other as adversaries. In most cases, for example, there are penalties for the general contractor if the deadlines are not achieved. On the other hand, there may be rewards if the project gets completed on time or earlier. In many cases these individual contractual objectives can cause an imbalance in the project. When the parties begin to see each other as adversaries, the blame game takes effect and can cause the flow of conflict poison even if the project is still technically on track.

If the negative energy is not addressed, it can create problems that are not even noticed at first. As advances are being made on a project the conflict poison will seem to be held at bay. The customer is happy that the project is moving forward. The contractor is happy with the cash flow, but the pressure of keeping things moving may not be allowing for the removal of the conflict poison. The contractor has many parties to deal with, many relationships to manage, not always allowing for proper "medication" to be applied to the project, causing an imbalance. The customer can feel this imbalance and may begin to wonder what the general is hiding. Trust becomes an issue and lack of trust is the most powerful conflict poison there is.

As other parties become involved in the project they bring with them their skill and expertise as well as their own experiences, which may contribute to the flow of conflict poison on this project. The "heart" of the project, namely the owner and general, must continue to check every party entering the project to ensure that the new blood is not already contaminated.

II. CONFLICT POISON — HOW IT STARTS AND GROWS

There are many types of disputes in the construction industry and parties involved in a project may have reasons for "fueling the flow of conflict poison", at times not even realizing they are doing it. Let us look at a handful of typical examples, starting with the perspective of a customer building a custom project.

When embarking upon a construction project, customers or owners usually spend an enormous amount of time and money ensuring that everything is in place to proceed with the project. Financing is in place and they have hired professionals, an architect and a contractor to manage the construction project. In addition, they have heard all the horror stories regarding cost overruns, project delays and work stoppages that can last for long periods of time.

Understandably, the customers embark upon the project with some nervousness because they have heard that even careful and experienced individuals run into problems during construction. Where those conflicts result in a work stoppage and lawsuits, costs begin to mount very quickly. The customers put a lot of trust in the construction team and hope that it will see the project to completion while protecting them from all of the pitfalls and problems.

The customers must understand that in almost all construction projects, there will be mistakes, misunderstandings and unforeseen problems. Constraints of time and money will pressure the customers into making decisions "on the fly", decisions that the customers may not be happy with later. Minor errors in building to specification may not harm the building, but it may change the aesthetics of the design, something the customers will not accept. The appropriate responses to these problems as they occur can either help the project, or they can increase the flow of conflict poison rapidly, bringing the project to a screeching stop.

The classic example of increasing the flow of conflict poison is a case where a cement company could not deliver cement on time due to a strike at the plant. When the strike was over and the foundation finally done, the carpenters were unavailable — they could not wait any longer, and started work on another job. This caused further delays. The customer had to pay extra interest on the construction loan and was not happy. He blamed the contractor and let him know how upset he was, saying they should have brought a non-union cement company in to provide the cement. The contractor took the customer's hostility toward him very personally, as he had worked long hours covering the footings and protecting the excavated area from the large amount of rain that had fallen during the delay. The contractor had also left the bulldozer on site to pull the cement trucks close enough to the hole to pour directly into the forms, an expensive decision he had made to save time and avoid the cost of a pump truck. The customer was too upset to recognize the extra effort and the cost savings provided by the contractor.

When a major problem develops like the one above, the customer may feel that the construction team has betrayed both the trust and working relationship. When the team starts to feel this conflict poison at work, things always get worse.

Another example of the way conflict poison starts and grows is when the weather plays havoc on the construction site. For example, in the morning, when the carpenters are supposed to commence the framing of the project, it starts to rain. After about an hour the carpenters leave the site to work indoors on another project. In the afternoon when the customer arrives at the site the sun is shining but the site is dead. The customer gets on the phone to his construction manager and asks why production is at a halt. The construction manager explains that the weather was bad in the morning so the carpenters left and will return the next day. The customer, not understanding how the industry works, angrily demands the return of the carpenters that same day. Once again the conflict poison begins to flow due to poor communication and a lack of understanding of the trade's situation.

A third example of conflict poison is when the product the customer selects looks different once it has been installed. The customer chooses a type of exterior brick, but upon seeing it on the house believes the wrong product has been used. He calls the contractor and asks for a meeting. At the site, the mason, the contractor, the brick supplier and the customer look at the brick. The supplier tells the customer that the picture he saw when ordering the brick was done using both a different color mortar and a rake joint. The customer is upset that he was not informed that mortar choices and styles would affect the final outcome, and is angry that his house does not look the way he intended. The "conflict poison" begins to flow due to poor communication and a lack of understanding of the options available.

In all of these examples, the relationship between the parties begins to fray and parties become distrustful and resentful, leading the parties to take steps they think will protect them from each other. The contractor or subcontractor, fearful that he or she will not be paid, may threaten to stop work and register a lien on the property. These threats, intended to protect the subcontractor's interests, will likely result in the owner feeling a need to protect his or her interests by talking to a lawyer. The lawyer, trained in the art of litigation, and feeling that the threats are unreasonable, may advise the customer to sue over the "deficiencies", leading to a complete breakdown in the relationship and the project.

III. DISPUTES IN THE CONSTRUCTION INDUSTRY

In order for a construction project to be completed effectively and on time, all parties must not only be singing the same song, they must also be singing the correct harmonies together. The conductor, usually the general contractor, must keep everyone on the same page, playing his or her part correctly at the right time. Each party relies on the other to do its job well and on time. Because everyone is so highly interdependent, the chances of delays, problems and disputes are very high.

Let us look at just how interdependent parties on a construction site are. For example, the carpenters need to build the forms for the footings. In order for them to do this, the general contractor must have the plans approved by the city. Next the contractor will notify the surveyor to stake out the excavation area to be dug and provide a benchmark for the depth cut. The footing materials need to be ordered 24 hours in advance to ensure they will be available on-site for the forming of the footings. Next the excavation company needs to be notified that the surveyor has completed the layout of the site. Then the excavation needs to be completed. Once that is done the surveyor must pin out the location of the footings in the excavation area. With the actual location of the footings determined, the carpenters can finally form the footings, but before the footings are actually finished, the city inspector must approve the work. Only now can the concrete be poured into the footings.

The general contractor plans a schedule to accommodate all the various issues well in advance of the job beginning. This schedule is a complex set of dependencies and interrelationships; any problem or change has a ripple effect that causes many problems for many parties. If the excavation company has a mechanical failure with its backhoe, it cannot dig the hole as per the schedule. A day is lost. The general contractor is now behind on the schedule, and must try to make up for lost time somewhere else on the project to prevent back charges and further delays.

(a) Different Types of Disputes

There are a number of different types of disputes that can occur on a construction project. There can be disputes between the customers, trades, manufacturers, suppliers, municipalities and service providers such as cable, phone, gas and hydro. Each dispute will have its own very unique environment and challenges. A conflict or delay in any one of these areas can jeopardize the entire construction schedule causing havoc to the entire project. Broadly speaking, some of the major types of disputes include timing, payment schedules, work quality, and change orders.

(i) Timing

Timing is the number one reason that disputes arise in the construction industry. The general contractor is responsible for the progress made in any project. Deadlines must be adhered to and delays caused by any party that impact this progress are not taken lightly. Every party involved in the construction project has agreed to the guidelines or the scope of work to be done prior to being awarded the contract. In the scope of work to be done there are stipulations such as defining the work, when the work will commence, what materials are to be used, what standard of quality must be met, how long will it take, and who is responsible for inspections, the payment schedule, warranty and insurance obligations.

Tenders initially go out from the general contractor to the subcontractors requesting bids. The subcontractors will get quotes and schedules from their suppliers, which will allow them to submit a bid with exact costing for their part of the project to the general contractor. Most parties involved in the bid process usually depend on some outside resources to establish this pricing and availability.

Now that the general contractor has received all the quotes back, an analysis is done to determine who will win the bid for each piece of work. Since the lowest bidder is often chosen, this creates a situation where profit margins are thin for everyone on the project. The general contractor will now submit the project tender to his or her customer, based on the bids received from the subcontractors. If the general contractor is awarded the job, he or she will expect that all of the subcontractor bids will be honoured. The performance and reputation of the general contractor is now on the line with the client, and there can be significant financial implications if things do not go according to the contract that has been executed.

(ii) Payment Schedule

Payment to individual parties for work completed is the next reason that disputes arise. Financial institutions provide most of the funding for construction projects. The funds are advanced as the work progresses. This is called a draw or "cost to complete". When a party issues an invoice, the party should be paid as stipulated in the executed contract. The party requesting payment needs to provide a statutory declaration, which states that everything provided to the project by them is or will be paid for; in other words, the party does not owe its suppliers any money. The owner or general is allowed to holdback up to 10 per cent of the payment for 45 days under the *Construction Lien Act*.[1] In many cases, however, the parties submitting their invoices ignore the contract payment schedule and expect payment sooner. This can cause hostility within the working relationship, sometimes leading to a work slowdown until payments are received. The parties involved take money issues very

---

[1] R.S.O. 1990, c. C.30, ss. 22 and 31.

seriously, and whenever there is conflict about the amount or timing of payments, the conflict poison begins to flow.

(iii) Substandard Quality

Quality of work is the next type of dispute that comes up time and time again. If the quality of work is not as per the contract, or the customer feels that the quality of work is unacceptable, the conflict poison rapidly begins to flow out of control. Trades are typically very proud people, and if their work quality is questioned, they take it very personally. Arguments about work quality will impact not only the construction schedule, but if repairs are needed, it will also cause a delay in payment. Quality issues erode trust, damaging the working relationship.

(iv) Change Orders

Changes or change orders are another main reason for disputes to occur. In custom projects the customer or designer can change things every day, frequently causing work to be scrapped and redone. When done properly, extras and changes must be accompanied by a signed change order that stipulates the cost of these changes. Time and time again, however, trades make changes, often at the urgent request of the general contractor, and costs are not discussed until an invoice is submitted. Unless approved by the general in advance, these extra costs are always challenged and frequently will not get paid, regardless of the fact that someone urgently requested them. After the work is done, customers frequently say that if they knew the costs associated with the change, they would not have requested it. Refusing to pay for work completed poisons any future relationship.

(b) The Structure of a Construction Project

Before being able to understand why disputes arise in the construction industry it is important to understand the overall process of a typical construction project. We will focus on residential construction as we look at the construction process.

(i) Blue Prints

The first step in virtually every construction project is the blue prints, the full specifications to be followed during construction. A designer or architect usually provides these blue prints. On the blue print it will stipulate the full structural design and specifications of everything to be built, and these specifications must be adhered to. It is like a recipe for a good cake; the recipe must be followed if you want the customer to like the final product. Usually the client has spent many hours with the designer or architect going over the exterior and interior features of the home. Once the blue prints are finalized, the architect or designer will submit them to the building department for approvals and a building permit. The project should never begin until a building permit is issued.

It would seem obvious that once the customer signs off the blue prints, all is well. But is it? A set of blue print designs is completely on paper, and many people cannot visualize what the actual *as built* product will look like. If there is any gap between what the customer imagines the blue prints will look like, and what the final product actually does looks like, it will cause major conflicts along the way among the customer, contractor and designer or architect. The contractor could build the project exactly as per plan but the customer may reject it because it is not what he or she thought it would look like. If this process takes place, it will accelerate the flow of conflict poison.

(ii) Communication and Co-ordination throughout the Project

For every project, the general contractor will be required to communicate and co-ordinate the project with a large number of parties. To get a sense of the scope of this task, the following is a partial list of parties that may have a role to play in the construction process:

• Client	• Windows and doors supplier
• Architect or designer	• Garage door company
• Municipality	• Drywall and insulation company
• Surveyor	
• Insurance company	• Interior door supplier
• Temporary site labourer	• Trim company
• Excavation and large equipment company	• Painting company
	• Broadloom company
• Foundation forming crews	• Ceramic tile company
• Steel company	• Marble and granite supplier
• Lumber company	• Door hardware supplier
• Trusses company	• Kitchen cabinets supplier

• Rough carpenters • Roof shingle company • Heating and air conditioning company • Electricians • Plumbers • Fireplace supplier • Brick masons • Brick suppliers • Stucco company • Aluminum company	• Bathroom vanities company • Counter top company • Basement concrete finishing company • Drains and sewer company • Deck company • Landscape company • Walkway crews • Driveway and asphalt company • Garbage bin company • Temporary power provider • Site toilet provider.

Based on the list above, you can see why excellent communication, co-ordination and timing are crucial. Disputes can easily develop due to the sheer number of parties involved, and the complexity of scheduling and arranging them.

(iii) Finishing and Closing a Project — The Construction Lien Act

Each contract for a project includes a payment schedule. This schedule is usually based on a draw system or cost to complete. The contract will state when the contractor will be able to submit for payments, and the request for payment must be authorized by the customer, the engineer, the project manager or any other designate agreed upon.

The final sign-off is usually (but not always) based on substantial performance and/or completion and is governed by the *Construction Lien Act*.[2] In relation to substantial performance and completion, the Act stipulates:

> When contract substantially performed
> 2. (1) For the purposes of this Act, a contract is substantially performed,
>   (a) when the improvement to be made under that contract or a substantial part thereof is ready for use or is being used for the purposes intended; and
>   (b) when the improvement to be made under that contract is capable of completion or, where there is a known defect, correction, at a cost of not more than,
>   (i) 3 per cent of the first $500,000 of the contract price,
>   (ii) 2 per cent of the next $500,000 of the contract price, and
>   (iii) 1 per cent of the balance of the contract price.

---

[2] R.S.O. 1990, c. C.30.

Idem
(2) For the purposes of this Act, where the improvement or a substantial part thereof is ready for use or is being used for the purposes intended and the remainder of the improvement cannot be completed expeditiously for reasons beyond the control of the contractor or, where the owner and the contractor agree not to complete the improvement expeditiously, the price of the services or materials remaining to be supplied and required to complete the improvement shall be deducted from the contract price in determining substantial performance.

When contract deemed completed
(3) For the purposes of this Act, a contract shall be deemed to be completed and services or materials shall be deemed to be last supplied to the improvement when the price of completion, correction of a known defect or last supply is not more than the lesser of,
   (a)   1 per cent of the contract price; and
   (b) $1,000.

These two definitions are important because when a contract is deemed to be substantially performed or completed, under the Act's rules, the contractor must be paid, subject to the holdback allowed in section 22.

For example, in order to receive a payment the following must be done. An invoice is issued along with a Workplace Safety Insurance Board (WSIB) certificate and Statutory Declaration to the engineer. The engineer will then issue a payment certificate. When the customer receives the payment certificate he or she will then issue the payment. This will happen for every invoice issued *except* for the final "holdback" invoice.

At the end of a project, the owner is entitled to withhold or "holdback" a portion of the money due to ensure that the contractor has paid the subcontractors and suppliers. The Act allows subcontractors and suppliers to file a lien against the owner's property if they have not been paid. This holdback is done to protect the owner from these liens. Regarding holdbacks, the Act stipulates:

Basic holdback
22. (1) Each payer upon a contract or subcontract under which a lien may arise shall retain a holdback equal to 10 per cent of the price of the services or materials as they are actually supplied under the contract or subcontract until all liens that may be claimed against the holdback have expired as provided in Part V, or have been satisfied, discharged or provided for under section 44 (payment into court).[3]

The final holdback invoice is the 10 per cent held back from each invoice as per the *Construction Lien Act*. The Act, in essence, gives suppliers and subcontractors 45 days to file a lien if they have not been

---

[3] *Ibid.*

paid, after which no claim can be made against the owner.[4] After the 45 days has expired, the general submits the holdback invoice along with the WSIB certificate, the Statutory Declaration, as well as a certificate of substantial completion and a certificate of publication. The engineer will then issue a payment certificate for these funds.

(c) Interest-based ADR

Conflict of any kind, at any point in the process can drive parties to legal options and remedies, threatening the willingness and ability of the parties to finish the project on time and on budget. Interest-based alternative dispute resolution[5] is a way to dissolve this "conflict poison" by reaching agreements that are fair to all parties in an atmosphere of co-operation and mutual respect. More importantly, ADR can help restore damaged relationships through co-operation while advancing the interests of the parties.

Because most civil lawsuits settle before trial, the statistics are in favour of a settlement. The question, though, is when and how effectively the parties settle their disputes. If the settlement takes place two years later on the court house steps when both parties have spent thousands in legal fees, chances are that the resolution will make both parties unhappy and the relationship will be beyond repair in the near future. The ADR process is not only more cost-effective than litigation, but is also more satisfying because settlements meet the needs and interests of the parties without being imposed by a court, and therefore the terms are more likely to be honoured.

Delays in construction projects are costly, and because of the technical complexity of construction lien law, trials of these cases can take several years before the legal rights of all parties are addressed. For this reason, construction disputes can often be dealt with early by looking at the parties' needs and interests and not strictly or exclusively at their legal rights.

During these discussions, the parties are encouraged to move away from taking positions based on their legal rights, to a discussion of their interests. This requires everyone at the negotiation table to avoid "trashing and bashing". By avoiding positional behaviour, both sides are able to examine and discuss important interests that will never be addressed by

---

[4] *Ibid.*

[5] ADR is a term that covers the full range of alternative processes from negotiation to mediation through arbitration. "Interest-based ADR" refers to the non-binding, voluntary processes of ADR, such as mediation, facilitation and conciliation.

the legal process, such as getting the project completed on time, minimizing costs to all parties, and preserving relationships in the industry. Since the process is typically off the record, parties can freely discuss the details of everyone's behaviour, the work done and the costs incurred. Through skillful questioning, at times with the parties facing each other and other times with the parties apart, the facilitator or mediator will begin to understand what actually precipitated the flow of "conflict poison".

Once the parties understand each other's needs and perspectives, the discussion can shift to a consideration of the options that will help the parties. Different issues for each of the parties will be voiced at the table, and these differences in perspective are normal and should be expected. It is by recognizing these differences, then looking for creative solutions that maximize everyone getting what they can, that the ADR process can help parties reach solutions that work.

Even if complete agreement is not reached during this process, the parties can still aim for partial resolutions that keep the project moving, especially when confronted by issues that affect the construction loan draws and cause scheduling delays.

IV. SUMMARY

The remainder of this book will look at the full range of ADR options, both interest-based and rights-based, as well as identify best practices for managing and resolving conflict before, during and after completion of a project.

While the ADR process takes both commitment and work from all the parties involved, the process has the potential to stem the flow of "conflict poison", minimize the damage to the working relationship, and end the expensive delays that are part and parcel of any conflict. The more effectively all parties can access and use interest-based ADR early, moving on to rights-based ADR only if necessary, the more time they will spend in construction and the less time they will spend in conflict.

2
DISPUTE RESOLUTION – AN OVERVIEW

To identify and choose better ways of resolving conflict, we first need to look at the basics, at how we typically resolve conflict now. In general, there are three distinct types of processes, or approaches, that we use to solve problems, and each type or approach has its own pros and cons.

The three types of processes we use are:
- Interest-based processes;
- Rights-based processes; and,
- Power-based processes.

I. INTEREST-BASED PROCESSES

Interest-based processes are processes that focus on understanding and meeting each parties' "interests", their wants, needs, fears, and concerns. Interest-based processes assume that the best solution is one that substantially meets the needs and concerns of both parties, so that both parties are willing to accept and support the solution.

(a) Example of Interest-based Approach

Suppose two homeowners decide to build a fence around their backyard, and tell their neighbour. The neighbour hates the idea of a fence and asks why, to which the owners say that they are getting two dogs, and need an enclosure. They discuss the problem in depth, the neighbour explaining the reasons he hates fences (they are tall, ugly, block the light, make the yard seem smaller), and the owners explaining the type of dogs they are getting (cocker spaniels, which are smallish dogs) and why they need the fence (as an enclosure, space to run, *etc*). In the end, they work out a fence design that works for both of them (low fence, lets light through, nice design elements), and the neighbour offers to pay for half of the fence that runs along the joint property line. Both walk away with a solution they can live with.

Characteristics of Interests-based processes: This is an approach that tries to reconcile or find a solution that meets the interests of the parties. Interest-based approaches are or tend to be more consensual, and succeed when both parties get enough of their interests met to agree on a solution.	**Type of outcome:** Win/Win. **Type of process:** Collaborative. **Decisions**: Made jointly by the parties. **Requires:** Listening and understanding the other's point of view. **Process examples**: Most types of negotiation, facilitation, mediation, joint problem solving, mutual gains bargaining, communicating wants, needs, fears, *etc.*

II. RIGHTS-BASED PROCESSES

Rights-based processes do not look at the parties' wants, needs or fears, but rather focus on defining and asserting what each party sees as its rights or its entitlement in the situation. Rights-based processes tend to be independent of what the other party's concerns are, and therefore little time is spent listening or understanding where the other party is coming from. Rights-based processes often require formal hearings and third party decision makers, and typically end up with one party winning and one party losing (at least initially).

(a) Example of Rights-based Approach

When the neighbour hears that the owners are planning to put a fence up, he decides that he must stop the owners at all costs rather than talk to them about it. The neighbour hires a lawyer, who files an injunction against the owners to stop any fence construction. The owners, angry at being sued, retain a lawyer to fight the injunction. They all end up one day in a courtroom, where they all tell their stories. The judge, after hearing the arguments and looking at the relevant law, issues a ruling that the owners have the right to build a fence, providing they abide by all building codes and bylaws. The owners leave happy, the neighbour leaves unhappy.

Characteristics of Rights-based processes: This is an approach that is characterized by parties asserting or focusing on the superiority of one party's rights over the rights of the other parties. Rights come	**Type of outcome:** Win/Lose. **Type of process:** Adversarial. **Decisions**: Made by a third party, such as a judge, arbitrator, *etc.*

from many sources, including laws, statutes, conventions, past practices and precedents, policies, contracts, *etc*. Rights-based processes tend to be adversarial, and focus on promoting one's own rights while minimizing and de-legitimizing the other parties' rights.	**Requires:** Asserting one's own point of view, and rebutting or dismissing another's point of view. **Process examples**: Litigation, arbitration, adjudication, policy, tribunal decisions, formal investigation, *etc*.

III. POWER-BASED PROCESSES

Power-based processes do not look at the parties' wants or needs, nor do they look at the parties' legitimate rights or entitlements; they look solely at what a party can get through direct force, coercion, or the application of whatever power is available to it. There are many forms of power, ranging from physical power and violence, to economic power, reputation power, knowledge power, and many more. Power-based processes focus on winning and getting what is wanted by defeating the other party.

(a) Example of Power-based Approach

After winning at court, the owners buy their materials and begin constructing their fence, but on the second day they find that some of the wood is missing, and they are forced to buy more. The following day, they find the neighbour left the sprinkler on all night, making the ground too wet to dig post holes. The day after that, they find the post holes they dug were filled back in, because they were a safety hazard, according to the neighbour. After getting the fence partly up, the next morning they find that the fence has fallen over because the fence posts had been cut. The neighbour says he knows nothing about the issue. The owner, finally gives up and builds a small dog run on the other side of their yard, but not before hiring a security firm to build a tall steel fence on the boundary with the neighbour, complete with a security camera to prevent vandalism. The neighbour ends up with exactly the fence he hated most, and the owners have spent two to three times the money to get this done.

Characteristics of power-based processes: This approach is characterized by parties bringing to bear all the resources they have at their disposal against the other party in an attempt to win. Typically, power-based processes are highly	**Type of outcome:** Lose/Lose. **Type of process:** Highly adversarial. **Decisions**: Made by each party based on anger, fear, frustration, or revenge. **Requires:** Applying resources and power

adversarial, and are sometimes applied in spite of the rights of the parties.	with the goal of either winning or destroying the other side. **Process examples**: Threats, intimidation, coercion, physical force or violence, strikes or lockouts, unilateral decision-making, or "self-help".

IV. INTERESTS, RIGHTS AND POWER

We often put these three processes, Interest, Rights and Power, on a ladder or stairs as in Figure 2.1.

Figure 2.1

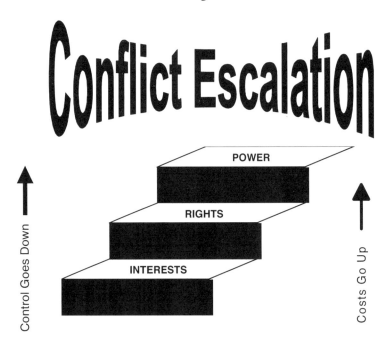

We do this because it illustrates two principles:

1. **Principle #1**: Interest-based processes (*i.e.*, negotiation, mediation, problem-solving, *etc.*) are lower cost, and more satisfying for the parties, than rights-based processes (*i.e.*, litigation, arbitration, *etc.*). In addition, rights-based processes are lower cost and more satisfying than power-based processes (*i.e.*, unilateral action, threats, intimidation, strikes, lockouts, self-help, *etc.*). The "best" (*i.e.*, the lowest cost and most satisfying) outcome is an interest-based resolution where the parties agree on a solution.

2. **Principle #2:** As you go up the stairs, control over the outcome decreases. When parties are using interest-based processes such as negotiation or mediation, the parties control the outcome. When it moves to rights-based processes, control moves to a third party such as a judge or arbitrator. This is significant, because even though we think of court as a place where each party has at least a 50/50 chance of winning, there are many situations where the outcome is lousy for both parties, and where both parties appeal the decision because they do not like it. When power-based processes are used, control is even less predictable. This is because while we can often assess who has more power, it is very hard to assess how much and how effectively either party will use its power. A wealthy developer with friends in City Hall may appear to have a lot of power, but if the environmentalists can get the attention of the press, or if they chain themselves to the bulldozers and prevent development in spite of the law, they may shift the power equation dramatically and unexpectedly. This is a regular occurrence in the world.

(a) Conflict Poison and Interests, Rights and Power

We have looked extensively at how conflict poison builds and flows on a construction project. In simple terms, conflict poison is when the business relationship deteriorates and parties begin to rely on rights-based processes, such as court, to get what they need, or they begin to apply power to force the other party to get what they want. When parties give up on the relationship, when they no longer trust each other or rely on each other to solve problems, the conflict poison begins. When one party feels it cannot rely on another party, or begins to assume that the other party will harm its interests intentionally or otherwise, parties tend to shift their focus to their rights and power in order to protect their interests. Focusing primarily on rights or power is a source of conflict poison in a relationship, and, left unchecked, it will only grow.

(i) *Barriers to Resolution: Why it is Difficult to Reverse the Flow of Poison*

Once a conflict has begun, once the poison is affecting the relationship, once trust is gone and parties see each other as adversaries, it can be extremely difficult to resolve the problem or to repair the relationship. Even if a quick resolution to the problem at hand is reached, if the relationship is poisoned a new problem is not far away. There are times when the parties themselves will find it extremely difficult to rebuild the relationship on their own. Indeed, research has identified a number of specific reasons why negotiation alone can fail to resolve issues or rebuild the working relationship.

Some of the major reasons parties fail to resolve issues on their own once the conflict poison has started are:

1. **Failure to Separate Substantive Issues from Relationship Problems:** Most issues in construction begin as substantive issues, meaning that a problem with the schedule, with the cash flow, or with the design of the building develops. At the same time, people have a strong need to blame someone for the problem, often triggering a defensive and angry response from the other side. Once this has happened, it becomes very hard for both parties to just focus on the scheduling problem or the equipment problem. Each time they try and discuss it, the anger and blaming returns, blocking or preventing calm discussion and problem solving. The negative relationship prevents solving the problem or triggers tit-for-tat behaviour, and the problem simply grows. Once a negative relationship develops it can trigger further problems, such as:

    (a) *Lack of Communication*: Once the relationship has soured, all parties restrict communication, or stop communicating entirely. They simply refuse to talk, making it impossible to find a solution to the problem. Frequently, if parties have consulted an attorney, they are advised to limit communications as much as possible, to prevent giving the other side ammunition for the coming lawsuit. As costs mount from the unsolved problems, even less communication takes place.

    (b) *Anchoring*: Once communication breaks down, parties become "anchored" to their own positions and their own views of what is right and fair. This anchoring is rarely based in fact, but rather is based on a party's own version of what has happened. As long as little or no communication takes place, the strength of the anchor for each side grows, entrenching the positions they hold.

    (c) *Lack of Trust*: Conflict tends to shatter and eliminate trust. Once trust is gone, so is any motivation to attempt a different solution on the grounds of, "Why bother, we can't trust them".

2. **Reactive Devaluation:** This is a fancy way of saying that once we are in conflict with someone, once we no longer trust them, we assume that everything the other party does is in some way bad and designed to hurt us. If the party reaches out with an offer to meet us halfway, we assume it is because he or she knows that he or she is 100 per cent wrong, they are trying to get the 50 per cent that they do not deserve. If the party apologizes, he or she is lying and do not mean it. If the party suggests a solution, he or she is trying to trick us. This mindset is very powerful, and causes us to assume that everything the other side does is dangerous.

The research shows us that the old phrase, "Seeing is believing" simply isn't true; we replace it with the opposite, "Believing is seeing". This means that if I believe you are untrustworthy and out to get me, everything you offer, even if it is something I said I want, I believe is really designed to harm me in some way.

3. **Hidden Agendas:** Once some or all of the above barriers are in play, both parties stop being honest about what they want and need, and begin to create and rely on hidden agendas. This is because we believe it is too risky to tell the other party, who we do not trust and who we believe is out to harm us, what we really want or need. We think he or she will simply use this against us, so we create new positions that will "throw him or her off the trail". Since the other side does not know what we want, or begins to assume we are always lying, it becomes impossible to find a solution that works for both of us.

4. **Fall Back to Positional Bargaining:** When faced with a situation where there is little or no trust, where ideas or suggestions from the other side are immediately discounted or ignored, where there is little open communication or belief that the other side is being honest with us, all that is left to do to try and solve the problem is old-style positional bargaining. This process consists of choosing an extreme position and demanding it be met by the other side, with few reasons given or expected. We choose an extreme position because we know we will have to give something up along the way, so we start by padding our offer to allow for this loss. The problem is, so does the other side, and we end up with numerous extreme offers thrown at each other, often through each party's lawyer. The whole process simply confirms for each party how unreasonable the other side is, and rarely leads to a solution that anyone feels is fair, let alone leading to any kind of ongoing working relationship.

These are significant barriers to resolving disputes in the construction industry. Next, we will look at the full range of ADR options and process, and how they can help address these barriers.

(b) Dispute Resolution Processes: This ain't the Stairway to Heaven!

Generally, when we look at best practices in dispute resolution, there is a sequencing of all ADR processes that can act as a guide. Similar to the Conflict Escalation stairs, the Dispute Resolution Stairway (Figure 2.2) is an expansion of those processes into more specific and identifiable steps. As readers (and especially Led Zeppelin fans) will see quickly for themselves, this is definitely not the Stairway to Heaven.

Figure 2.2

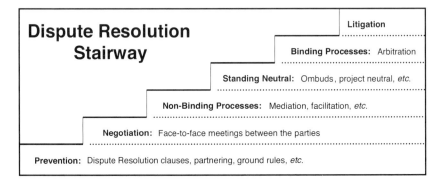

Let us take a look at each of the steps in Figure 2.2.

(i) Prevention

Prevention can take many forms, and there are many steps parties can take to anticipate and prevent conflict on the job site. The single biggest mistake parties make is in focusing prevention solely on rights- or power-based steps, such as penalty clauses in the contract or drafting 200-page contracts to try and capture every possible dispute and define fault. Good prevention steps focus on interest-based outcomes, and include the following:

- *Partnering*:[1] Partnering is a structured way for all key parties on a project to meet and commit to how the project will be run. Partnering does not change or affect the contract documents, but helps define and manage expectations and relationships throughout the project.
- *Establishing job-specific ground rules*: This can be done through Partnering, or separately and informally.
- *Dispute resolution clauses*: This refers to inserting dispute resolution clauses in the original contract that require parties to effectively use interest-based approaches if a dispute arises.
- *Appropriate allocation of risk*: There is a tendency on the part of owners to try and prevent conflict by writing a contract which shifts most or all risk to the contractor, whether that risk is indeed reasonable for the contractor to assume or not. Two things happen in this case. First, the cost of the contract tends to go up, at times significantly. If a contractor is to assume risk of all kinds on the project, even risk that he or she does not control, a contractor will tend to pad the contract to try

---

[1] Partnering will be looked at in depth in Chapter 3.

to account for that. Second, with an unfair or unreasonable allocation of risk, the relationship is adversarial from the beginning, increasing the likelihood of conflict and decreasing the likelihood of resolution. Risk should be clearly and reasonably allocated from the beginning of the process.

(ii) Negotiation

It may sound obvious, but negotiation should be the first step in any resolution process, because it is fast and cheap. It must, however, be done in a way that will lead to resolution and problem solving, not lead to positional bargaining that will escalate the conflict. Negotiating through email, voice mail or snail mail through counsel will tend to escalate rather than resolve the conflict. In addition, the ground rules for the negotiation should be clear, *i.e.*, the goal should be to solve the problem, not shift the blame to the other party.

(iii) Non-Binding Processes

This step refers to a whole range of processes that involve the help of a neutral third party who does not impose a binding solution on the parties against their will. These processes include:

- *Facilitation*: Facilitation can be done by just about anyone trained and able to keep the parties focused on effective problem solving, including someone involved in the project as one of the parties. Facilitation is less formal and structured than mediation, and works best when (1) the level of conflict is low to medium (2) most of the issues are substantive, and (3) the parties still have a reasonable working relationship.
- *Mediation*: Mediation involves a trained, professional neutral who is not related to any of the parties to the dispute and who has knowledge about the construction industry. Mediation is more structured than facilitation and is used when the dispute has escalated. Mediation is often used when there are not only substantive issues to be resolved, but also significant relationship and communication issues that prevent resolution.

(iv) Standing Neutral

This is more formal than facilitation or even mediation, and usually involves the help of someone appointed ahead of time as a neutral the parties will use in conflict situations. Ombudspeople are the best example of this. Larger construction projects have begun to appoint a person on retainer as the official standing neutral that any party can call on to help if a conflict arises. A standing neutral can also go beyond the informality of mediation or facilitation by providing a Neutral Evaluation, an opinion on who is right or what a reasonable solution might be, and while this opinion may not be binding on the parties, it often carries significant weight to help move the parties toward resolution.

(v) Binding Processes

The most typical binding process parties use is arbitration. This is where a neutral third party hears a dispute, often running the hearing in a similar fashion to a courtroom, and ends the process by issuing a decision that is final and binding on all parties. Typically, there is no appeal from the arbitrator's decision. Arbitration is frequently seen as a faster, cheaper way than litigation to get a binding decision, but because there is no appeal of the arbitrator's decision, it is sometimes seen as a risky choice. In addition, arbitration can be just as expensive as litigation in many cases.

(vi) Litigation

Litigation is well known to parties in the construction field, especially for the time it takes to go through, for the costs incurred by all parties, for the uncertainty of the outcome, and for the length of time any appeal processes take.

(c) Interest-based Processes Reverse the Flow of Conflict Poison

Effective use of ADR can help keep parties low on the Dispute Resolution Stairway, and focused on what they are there to do, *i.e.*, complete the project. So how do interest-based processes and ADR help prevent the conflict poison from killing the project? There are a number of ways.

- *ADR Re-focuses Parties Back to their Important Interests*: Interests are our motivators; they are at the core of what we do and why we do it. Rights and power are simply other ways (and often neither effective nor satisfying ways) for us to try to get what we want. By re-focusing the parties' attention back to interests, to what everyone really wants from the construction project (rather than on things like revenge, which

parties want because they are not getting what they *really* want), everyone can begin to take actions that help the project move forward.
- *Focus Away from Negative Relationship Desires*: We all carry a desire to repay the way we have been treated. If we feel we have been attacked, we tend to attack back in some way, shape or form. By focusing on constructive interests, parties can channel the energy and emotion away from revenge and payback and toward the original goals of the project. This channelling focuses parties on moving forward, and can transform the entire project.
- *Slowly Build or Rebuild Trust*: By focusing on constructive steps, the conflict poison will slowly drain away, and in its place small amounts of trust will begin to grow. Over time, this can recreate or rebuild damaged business or working relationships.

More specifically, interest-based ADR processes such as mediation and facilitation can be extremely effective in overcoming the barriers we looked at earlier. For example:

1. **Failure to Separate Substantive Issues from Relationship Problems:** Mediation and facilitation are both effective at helping parties to clearly define the different issues that need to be resolved. This includes identifying what are substantive, construction-related problems, and what are relationship or personality type problems. Each of these problems will need a different type of approach, and mediation can keep that clear by focusing the parties on solving these issues one at a time. For example, substantive problems frequently need more data, and sometimes need an expert or specialist to help the parties find a solution. Relationship issues must often allow venting, encourage listening and allow apologies, to be effectively resolved. Mediators are trained to help separate these issues, and focus the parties on what they need regardless of the type of problem they are facing. Specifically, interest-based ADR can assist in the following way:

    (a) Lack of Communication: Mediators are trained to manage communications effectively. Mediators run the process so that each party has an opportunity to have its say and be heard by the other parties. Many issues can be resolved simply by creating a forum for effective communications.

    (b) Anchoring: Mediation and other interest-based approaches create safe and open discussions. This allows mediators and facilitators to effectively question each party's anchors, and challenge each party's "beliefs", often helping parties to see the problem differently.

    (c) Lack of Trust: Mediation and facilitation are excellent for bridging a lack of trust between the parties, because the parties can initially put their trust not in the other party, but in the mediator and

the process the mediator is running. By doing this, the lack of trust barrier can be minimized, helping the parties to get back to work on solving the issues. Over time this will help the parties rebuild trust in each other.
2. **Reactive Devaluation:** When reactive devaluation is at work, each party discounts and demonizes the other party's behaviour. By having a third party filtering and framing the entire process, the process itself has more credibility. In simple terms, the mediator can make suggestions or put out ideas for the parties to consider without triggering the devaluation of those ideas, because they come from the mediator and not from the other side.
3. **Hidden Agendas:** Mediators are trained at challenging and questioning each party's interests, and will frequently hear from each party in caucus what they really want, as opposed to what they may have been saying. Mediators can guide each party to communicate directly what they want and need, effectively eliminating hidden agendas along the way. In cases where a party misleads the mediator, aligning what a party says it wants with clear ways for it to get it, a mediator can often "flush out" hidden agendas effectively.
4. **Fall Back to Positional Bargaining:** Parties frequently know that positional bargaining will simply not be effective in getting a good solution. By getting parties focused away from their positions and more onto their interests, the negotiations stop being as positional and become a problem solving exercise instead.

V. SUMMARY

This chapter has provided an overview of some of the ADR and interest-based options that parties can use to get faster, cheaper and better solutions to disputes. In the next three chapters, we will focus on best practices in the construction field for applying ADR before the project starts, how to apply ADR during a project when disputes arise, and how to apply ADR after the project if disputes still exist.

3

BEST ADR PRACTICES — PREVENTION AND EARLY DISPUTE RESOLUTION

I. PREVENTION — THE BEST MEDICINE

There are a number of areas that contractors and clients should pay close attention to before any contract is signed or any project is even considered. These areas range from assessing the other party on both a personal and a professional level, to structuring the contract effectively.

(a) Align Expectations — Be Comfortable with Who you Hire

In some ways, construction projects are arranged in reverse of good business approaches. Typically, a project is bid (sometimes without ever meeting the key people from the company doing the bidding), the lowest bidder is accepted and contracts are signed, and only then do the parties even think about how they will accomplish the work together. While this may appear to be good business by getting the lowest cost to build, it can quickly turn into the highest cost decision if conflict and delays take place due to poor communication and poor relationships.

Before getting into the specifics of the project itself, get comfortable with the other party. Spend some time talking about past jobs, common issues and how they would be dealt with, significant concerns the contractor has, significant concerns the client has, and how they would be handled or addressed. This is not small talk or socializing; it is an important part of aligning the parties' expectations. Some clients, for example, want very little information other than schedule and budget updates every few weeks. Other clients require more "hand-holding" or reassurance that the project is on track, and want detailed descriptions every few days. Neither approach is right or wrong, but both the client and the contractor should be on the same page. Unless the expectations of both parties are clear and aligned, conflict will result. Once the contracts have been signed, compatibility issues are harder to deal with.

Specifically, parties should address and agree on the following areas:

- Communication needs: who needs what information, and why;
- Frequency of meetings, type of information;
- Change requests;
- Relative priorities: schedule, cost, aesthetics, *etc.*;
- Who can talk with subtrades;
- Other expectations: site cleanliness, variances, *etc.*

The contractor should find out what the client needs prior to beginning, perhaps even prior to bidding the job, if possible. If an extra one or two meetings per week with the client are required for the client to be happy, this needs to be built into the process rather than avoided.

While some of these issues will be directly addressed in the contract, do not assume that is enough. The interpretation of contract clauses has generated a great deal of court time, so it is better to discuss these issues up front before signing the contract to know if the parties are a good match.

(b) Detailed Discussions about the Quotes

There appears to be a lot of reluctance to discuss the different bids received with the contractors who are doing the bidding. Different quotes will treat different aspects of the job in different ways, and underlying all quotes are a series of assumptions about how each particular piece of work will be accomplished. It is nowhere near enough to say that certain items will be built to code, because the building code leaves significant room for interpretation too.

The solution is to have detailed discussions with the contractors and their bids. Ask enough questions to be sure you are comparing apples to apples, that similar-looking items on two different bids are actually the same. Some examples of discussions that should take place include:

- Are the materials being used in the different bids the same? What are the quality differences? For example, are the roof shingles of the same quality? What underlays will be used?
- What preparation or remedial work will be done before construction? For example, what thickness of gravel will be laid before pouring the concrete? How will it be compacted?
- What is the cost if additional work needs to be done? How will that be priced and agreed?
- What brand and quality of paint will be used? How many coats? How will colour samples be verified?

This is the time to put all the cards on the table, to go into some detail about what expectations each party has, and what each party's priorities are. At the end of this discussion all the parties should have a good understanding of the scope and cost of work to be done. If parties do not get along during these initial discussions, if it is not comfortable for parties to discuss these issues up front, then the parties are probably setting themselves up for disputes down the road.

(c) Check out References

When references are provided by the contractor prior to signing contracts, they should be investigated. It should be no surprise that the references provided will probably be strong, but check them out anyway. Another way of asking for references is to ask the contractor for a list of his or her last 10 jobs, from which the client can pick any two to call. This is a way of getting a better idea of how the previous clients feel they have been treated. Avoid references from projects currently underway, as the clients may have an interest in keeping the contractor happy in order to get their own job done. If there is any opposition to contacting any of the provided or requested references, this should be a major cause of concern.

When talking to references, go beyond the question of whether they would recommend the contractor or not. Remember, that particular client may have very different needs and expectations than you do. Ask detailed questions, such as:

- How often did the contractor charge extras, and how fair and reasonable did you think that was?
- Did the contractor complete the project on time? If not, why not?
- How easy was the contractor to work with? What are some examples?
- Was the contractor available and easy to communicate with? How frequently did the client meet with the contractor, what kind of meetings did they have, and how detailed were the status reports?
- Were the subtrades the contractor used easily accessible and professional?
- Was the work done in a professional manner?
- How clean was the site kept?
- How responsive was the contractor to your concerns? What is an example of that?
- Could you look at the project the contractor did for the client?

Another source of information on the contractor is the building department where the building permit will be issued from. Call the department, and ask for the following information:

- Have you ever dealt with this contractor?
- Are there any outstanding issues with this contractor?
- Do you know of any problems with this contractor?

While the people in the building department are not required to give you this information, in many cases they will be happy to give you some feedback.

Lastly, the contractor should ask for and check the financial references of the client to ensure that the client has the ability to meet funding deadlines, minimizing the possibility of liens on the project. Financial questions that should be asked of the client include:

- How is the project being funded? Is there a construction loan, from whom, and what are the requirements for release of funds? Ask if you can talk with the lender to validate this information.
- Do a search on the property to ensure there are no liens or outstanding orders already in place. If everything is clean it will ensure commencement of the project.
- Talk to the architect or engineer who drew the plans and ask questions about the payment history of the client, accessibility, and responsiveness of the client if issues arise.

Remember that the lowest price is not always the cheapest in the end, and by doing the investigation discussed you will be able to minimize the chances of finding this out too late. Doing a complete check and getting everyone on the same page will help to ensure a successful relationship.

(d) Partnering

On larger construction projects, partnering is one of the most effective prevention measures that can be taken.

Partnering is a preventative dispute resolution process that developed in the United States construction industry approximately 20 years ago. It developed in response to high levels of conflict and a large volume of legal claims on major construction projects, and was designed to be a better way to manage conflict and promote collaboration between parties working toward a common goal.

Partnering has, for the most part, been an unqualified success, resulting in widespread use on mid-to-large construction projects in many countries in the Western world. The research that has been published on the partnering process has been clear and unequivocal. Partnering works. Period.

To give an idea of how effective partnering is, one research study that highlights the success of Partnering comes from the Arizona Department of Transportation (ADOT), who adopted Partnering on all of its projects in 1992. The previous five years of claims data for ADOT looked like this:

1987 Claims on ADOT projects against ADOT	$ 16.5 million
1988 Claims on ADOT projects against ADOT	$ 21 million
1989 Claims on ADOT projects against ADOT	$ 21.5 million
1990 Claims on ADOT projects against ADOT	$ 23 million
1991 Claims on ADOT projects against ADOT	$ 25 million

In 1992, all projects were Partnered, with the following results:

> The results of Partnering speak for themselves. During the first two years of Partnering in Arizona, **there has not been a single new claim or issue which has not been resolved** through the issue escalation process. ADOT changed from an organization where great effort went into claims resolution to one where it no longer needed to invest manpower in this way. (emphasis added)[1]

In relation to completion time on projects, in fiscal 1991, 27 per cent of all projects failed to be completed in the original contract time. Through Partnering, this was reduced by over 74 per cent, with only 7 per cent of projects in 1992/93 completing after the original contract time. Many projects actually began completing early. As the report stated, "No longer is one in four projects completed after the original end of contract time: rather, the question now is how much ahead of schedule the project will be finished."[2]

For example, one project scheduled for 13.5 months completed in 8.5 months. Another was set for 16.5 months, and completed in 8 months.

Finally, in relation to how well projects hit their projected budget, prior to Partnering the historical increase or overage (including contingency) was an average of 5 per cent above the contract price. That has now fallen to 3 per cent as a new trend. This means that the State of Arizona began saving a full 2 per cent of the total contract cost each and every year; this represented major savings for the State, and was valued well into the millions of dollars per year.

(i) What is Partnering?

So, what is Partnering? Essentially, Partnering is a one- or two-day workshop conducted with all the key players on a construction project after the contracts are bid and signed, but before the project actually begins. Essentially, the time is spent focused not on what the participants

---

[1] T.R. Warne, *Partnering for Success* (New York: ASCE Press, 1994), at 55.
[2] T.R. Warne, *Partnering for Success* (New York: ASCE Press, 1994), at 57.

are going to build, since that is spelled out in the designs and the contracts, but rather focused on *how* the parties are going to build it. It focuses on core principles of alignment, communication, decision-making and issue resolution. And by doing this, it dramatically changes how parties behave on a construction project.

In other words, parties need to commit to putting a significant focus on the co-operative aspect of their relationship, and Partnering is effective at helping them create this focus. Their overarching goal of mutual success must be in alignment and given strong focus.

(ii) The Partnering Workshop

The Partnering workshop is the formal intervention that helps structure the Partnered relationship. To this end, the activities of the workshop fall under one of four steps, as follows:

1. **The Partnering Charter:** This is where the participants develop and agree to a mission/vision statement; either a logo, a motto, or both; and a specific number of measurable goal/objective statements. Functionally, what the charter exercise does is help achieve the alignment of all parties' goals, objectives and values on the project, with all the key participants from each organization present. Since everyone works together to make the charter, the process also contributes to forming new relationships and building teamwork among the various people on the project.
2. **Roles, Challenges and Opportunities:** Every project has both unique challenges and problems that will need to be addressed, as well as unique opportunities that the parties can take advantage of. During the workshop, cross-membership teams identify and clarify the specific challenges, the opportunities, and people's roles on the project, after which they then develop and brainstorm solutions and ideas specifically to address these challenges.
3. **Issue Resolution Ladder:** This is a facilitated design process where the participants develop and commit to a focused, structured and rapid process for addressing and resolving problems, with the goal of resolving most problems at the front line, avoiding the delays that formal steps such as liens and litigation often bring. Note that this does not take away anyone's rights to access the courts or arbitration to resolve disputes; it simply creates a more immediate, interest-based forum for finding solutions quickly and effectively.
4. **Legacy Structures:** No matter how effective a one- or two-day workshop is, the workshop achievements pale when compared to a one-year, two-year or ongoing long-term relationship filled with challenges and problems that need solving. This section helps the

parties to design and agree to support structures that maintain the focus on strong collaborative approaches to problem solving long after the workshop has ended. To this end, Partnering "champions" are nominated from each party who will keep the Partnered approach in focus by encouraging the resolution of any issues that are impairing the quality of the relationship.

Essentially, then, Partnering is a mutually developed, "formal" strategy for creating and nurturing good communication and high levels of commitment between various parties working together on a project. While the idea of good communication and strong commitment between parties on a project is certainly not new, the way that it is accomplished and implemented through Partnering seems to be of great value in achieving good outcomes for the parties.

Partnering is a tactical methodology for building and maintaining strong and effective relationships on a project. Notice that Partnering is silent on the strategic issues implicit in any project, *i.e.*, *who* anyone should Partner with. Partnering is not a process to help parties decide who the best partner on a project will be, but rather is a methodology to assist the parties once they have chosen or decided who will build the project.

(e) Issue-resolution Process

There are two aspects to Issue Resolution, the formal and the informal processes. The formal processes should be built into the contract itself in the form of dispute resolution clauses in the contract documents, and this is addressed below in the Key Contract Provisions section. The informal processes are agreements between the parties as to what will happen *before* anyone triggers the formal processes in the contract (as noted above, Partnering is very effective at creating the informal issue resolution process). Both are important.

When issues arise during the construction project, it is important there is an informal method in place that allows the parties to express their individual concerns without escalating the conflict or offending the other party. Strong relationships are critical to helping the parties resolve issues as they arise. When an informal but structured procedure is in place that all parties have committed to using, it will make it easy for the parties to address difficult issues effectively. So how is this done? A number of steps should be clearly laid out, including:

- Open communication through regularly scheduled progress meetings, and a commitment from all parties to raise issues quickly;
- A clear procedure for raising and addressing emergency issues between regularly scheduled meetings;

- A commitment to meet within agreed timeframes, and to define and understand the issues before decisions are made. In other words, no one will dismiss issues out of hand without jointly looking at the information and understanding the problem from everyone's point of view;
- Agreement that all issues will be documented and decisions recorded with deadlines and responsibilities noted.

When an informal issue resolution process is followed it can help to stop disputes from escalating, because the parties involved are committed to tackling issues early and working together to resolve them. For further ideas on this, note the Issue Resolution section under Partnering in this chapter.

(f) Communicating vs. Decision Making

Another key prevention issue is distinguishing between communicating and decision making. In most cases, because the construction process is so complex, parties control the process by limiting or restricting communications. The owner will refuse to talk to the contractor or the trades, directing them to the consultant. The contractor will order all trades to avoid talking to the owner or the architect, and direct them back to the contractor. The net result of this is slower decision making and poorer communications.

Separate these functions. Free up everyone to talk to whomever they need to to ensure that important communication takes place. In addition, make it crystal clear that once the communication is over, any and all decisions made must go through the proper channels. For example, if the mechanical company wants to know if it can move the HVAC equipment on the roof 10 feet north to accommodate a roof vent, it can and should be able to talk to the engineer who designed the roof and confirm that the roof will support the weight in the new location. That is communication, and it is important because if the move is not acceptable to the engineer, the mechanics should not waste any of their time going down that path. If the move is okay with the engineer, however, the mechanics must now get a decision for the change, and to do that, they must go to the general contractor for approval. This step is important because the contractor must ensure that nothing else will be impacted by the change, including the schedule.

By making the distinction between communicating and deciding clear, maximum communication can take place without changing the responsibilities of the parties.

(g) Key Contract Provisions

When working together on any project, the first step is to create a foundation upon which job performance will be based. After the initial checks on compatibility and alignment have been completed to the parties' satisfaction, it is now time to move on to the next step. A written contract must be executed. This will be the legal bond between the parties involved in the project.

Contracts, of course, are legal agreements between the parties, and good legal advice should be obtained to ensure that the contract covers all the legal issues that need to be addressed. It should also be noted that contracts should go beyond the legal issues themselves, and should address the key interests of the parties as well. Below are a number of areas that should be addressed by the contract.

(i) Who is the Contract Between?

This sounds simple and obvious, but may not be. The parties privy to the contract should be clearly spelled out, but most importantly should be the same parties that the compatibility, relationship building and reference checks were conducted with. If the contract is not between these specific parties, the reasons should be investigated. There are a number of reasons why the legal agreements may be with different parties than expected, some of these reasons valid and some not. Examples of concerns about this include:

- Resources and reliability may be different that expected. If the individual the contractor has been dealing with has strong financial resources but the contract is in a corporate name, those resources may be an illusion when problems arise. Find out what resources the contracting entity has, and base your decisions on that.
- The decision makers might not be at the table yet. It may turn out that the person you think you will be dealing with has no authority from a legal, contractual point of view. Find out who has the authority, and meet with them.
- Liability and responsibility can be significantly affected. While the person you deal with may make promises, if they are not the contracting entity you may have a hard time holding them accountable. Find out about this up front.

The parties may have quite legitimate reasons for using different names on a contract, but these areas should be explored and questioned prior to execution of the contract.

(ii) What is the Address of the Project?

You may have heard the story of a building that was built on the wrong property. It can and has happened. Indeed, recently on the home renovation television show "Facelift" with Debbie Travis, her team got halfway through a renovation before finding out they were in the house next door to their client's. This occurred despite best efforts of an entire television production crew. Ensure that the correct address of the project is on the contract and further, and go to that address together so there is no possibility of error.

(iii) Dispute Resolution Clauses

Dispute resolution clauses should always be put into the contract, without exception. Dispute resolution clauses are nothing new to the construction field. The Canadian Construction Documents Committee (CCDC) is a national joint committee in Canada whose membership includes one owner's representative each from the public and private sectors, along with one representative each from the Canadian Construction Association (CCA), the Association of Consulting Engineers of Canada (ACEC), Construction Specifications Canada (CSC), and the Royal Architectural Institute of Canada (RAIC). In addition, one lawyer from the Canadian Bar Association, construction section, sits as an *ex officio* member. The CCDC has produced standard draft contracts for the construction industry for many years now, and its contracts have a wide range of dispute resolution mechanisms to consider.[3] The CCA also provides draft standard contracts for parties to use.

For example, the CCDC 4 Unit Price Contract has little in the way of formal dispute resolution. In this contract, a "consultant" who is appointed by the owner and is the owner's representative during construction shall be the "interpreter of the requirements of the Contract Documents and the judge of the performance" if there are disagreements. In spite of working for the owner, the consultant is not supposed to "show partiality to either party". There are few contractors, however, who will accept a decision they do not like from a consultant who is paid by the owner.

Two other contracts, the CCDC 2 Stipulated Price Contract, and the CCA 1 Stipulated Price Subcontract, have more detailed dispute resolution clauses. In both cases, the contract requires parties to follow these steps:

---

[3] See Appendix 4 for a reprint of the full dispute resolution clause from the CCDC 2 contract.

- *Step One*: the party who does not like the consultant's decision is directed to put the claim in writing detailing the particulars and give it to the other side. The other side is required to respond in writing addressing the issues. All of this is to ensure that the dispute is clear to each party, and that both parties know where the other stands.
- *Step Two*: If Step One does not resolve the problem, parties are then required to negotiate in good faith to try to resolve it. The contract makes such negotiations off the record, or "without prejudice", to encourage frank and open discussions.
- *Step Three*: If negotiation does not resolve the problem, parties are to jointly request the project mediator to mediate the dispute.
- *Step Four*: If mediation does not solve the problem, either party has 10 days to request binding arbitration; if requested within 10 days by one party, the other party is bound to go to arbitration. If neither party requests arbitration, then and only then can parties access the courts.

Note a few things. First, all the above steps take place within very tight timeframes, typically 10 to 15 working days for each step. This is to prevent any party from causing excessive delays. Second, all of the contracts mentioned include clauses that require the contractor or the subcontractor to continue to perform the disputed work during the resolution process, again as an attempt to prevent delays that will create more claims and greater costs (note also that the contracts make clear that doing the work will not prejudice any party in any way). Finally, while some contracts make mediation a mandatory part of the dispute resolution process and other contracts allow parties to skip mediation if a project mediator is not appointed, in all cases either party may withdraw and terminate the mediation at any time, reinforcing the idea that mediation is essentially a voluntary process for the parties.

The key provisions that parties must consider in a dispute resolution clause are the following:

- What steps will be mandatory and cannot be avoided?
- Who will function as the neutral, whether mediator or arbitrator, and how will that person be chosen?
- What work must continue during the dispute resolution process?

In terms of what steps are mandatory, good dispute resolution clauses make both negotiation and mediation mandatory. What is mandatory in negotiation and mediation is that parties attempt resolution, since any agreement reached must be voluntary and acceptable to both parties. Since negotiation and mediation do not allow anyone to mandate a solution over one party's objections, they are considered reasonable processes to require the parties to try. In other words, mandatory negotiation and mediation are really ways of creating opportunities for the parties to sit down off the record, while the conflict poison is still relatively low, and fix the problem. By making the sessions legally "without prejudice" it protects all parties' rights, again making it a safe thing to do.

The harder decision is whether to make arbitration mandatory. Since arbitration is final and binding and parties have very limited ability to appeal arbitrated decisions, it can be seen as a risky process. The options are to leave it out and make the courts, with their appeal mechanisms, the final step for parties, or to make arbitration mandatory if either party requests it (as CCDC and CCA have done). Parties should carefully consider the pros and cons before making this decision (see Chapter 5 for more information on arbitration).

Finally, as to how the mediator or arbitrator is chosen, there are two ways. The neutral can be appointed in the contract by name, if parties agree to that ahead of time. This is what a project mediator or standing neutral is, as described below. Another alternative is to appoint a professional ADR association as the organization who will name the neutral or run the appointment process, and identify that organization in the contract. Below is a sample clause from the ADR Institute of Canada, a non-profit organization who appoints neutrals on behalf of parties. Note that this is a generic clause that can be inserted as written, or can adapted for any particular contract.

> All disputes arising out of or in connection with this agreement, or in respect of any legal relationship associated with or derived from this agreement, shall first be mediated pursuant to the National Mediation Rules of the ADR Institute of Canada Inc. All disputes remaining unsettled after mediation shall be arbitrated and finally resolved pursuant to the National Arbitration Rules of the ADR Institute of Canada, Inc. The place of mediation or arbitration shall be [specify City and Province of Canada]. The language of the mediation or arbitration shall be English or French [specify language].[4]

Note that there is reference to the mediation and arbitration rules of the ADR Institute of Canada. Those rules contain clear guidelines for how the dispute shall be administered and how the neutral shall be selected or appointed. By adopting the rules as such, it simplifies the contract clause and builds in a strong process administered by a neutral organization.

Regardless of the final format, parties should have a clear and specific clause in the contract that requires parties to meet and attempt to resolve disputes quickly, before the conflict poison ends any chances of reasonable resolution.

---

[4] ADR Institute of Canada National Arbitration Rules.

(iv) Standing Neutral

On larger projects, appointing a standing mediator or dispute resolution neutral is an effective mechanism, with the goal of resolving disagreements and disputes as soon as they occur. This standing neutral can intervene early, preserve the relationships, and prevent the project from coming to a halt.

There are a number of roles that the standing neutral can play. Initially the neutral will facilitate communication between parties, help clarify positions, encourage the parties to keep the project moving forward as they work on the issues, thus saving time and money for everyone. To accomplish this, the neutral is most effective if involved from the conception of the contract to ensure that the neutral clearly understands what all the parties initially agreed to.

One of the major problems with construction disputes is that once the conflict poison begins to flow the relationship between the parties involved deteriorates very rapidly. Everyone involved goes into "CYA" (Cover Your Assets) mode, which typically means "papering" the file with letters and demands, avoiding decisions that may create liability, and communicating as little as possible to avoid divulging information that might cost money later on. This is the major reason for all work to stop when a problem arises. As discussed earlier, once there is a delay or slowdown in the construction process, it causes a snowball effect. The scheduling is disrupted, money stops flowing, and it takes great effort to get the project back on track. If not resolved, the conflict poison will cause further delays and costs, increasing everyone's commitment to blaming others in an effort to avoid the growing cost and liability.

The standing neutral can cause the parties to deal with the problem when it arises, avoiding the snowball effect by helping resolve the issues early with minimum disruption. The neutral must be someone on call 24 hours per day, 7 days per week, and must have an in-depth understanding of the construction process and construction issues.

Essentially, there are two roles the standing neutral can play, depending on the desires of the parties at the start. First, the neutral can simply be a mediator, someone who brings the parties together and helps them to find reasonable solutions they can agree on. Second, the standing neutral may be asked to arbitrate or decide certain issues and make the decisions binding on the parties. In some cases, the neutral can only decide issues of a certain magnitude, ending the deadlock and getting the work going again. In other cases, the neutral is only allowed to make temporary decisions, ones that are implemented today to keep the work going, but can be challenged in court by any party at a later date. This approach keeps the project moving, minimizes the damages delays can cost, while preserving the right to appeal the neutral's decision in another forum.

Whatever the role of the neutral, it must be clearly spelled out in the contract to the satisfaction of the parties.

(v) Risk Allocation

One of the main functions of a contract is to allocate risk between the parties, yet it is not uncommon that the contract between the owner and the contractor fails to do this effectively. This process typically fails along one of two lines: either the risk allocation is left undefined and unclear leading to disputes over who was responsible for what risk; or the owner shifts all of the risk to the contractor. This shifting of the risk, often called "risk management", typically refers to the inclusion of "disclaimer clauses" into the contract. These contract terms are focused on excluding the owner's risk and liability, both in contract and often in tort, and making it completely the contractor's risk and liability. In many cases, this risk is shifted even though the contractor may have no means with which to control the given risk (such as weather, soil conditions, delays due to strikes, *etc*.)

There are two significant problems that come out of poor risk allocation. First, the costs of the contract can go up. In general, these costs have been estimated to be in the 8 per cent to 20 per cent range, meaning that the contractor adds between 8 per cent and 20 per cent to the contract price to account for the increased (and possibly inappropriate) risk that he or she is taking on.[5] More importantly, perhaps, is what poor risk allocation does to the contractor/owner relationship. In general, this shifting of risk reduces the level of trust between the owner and contractor, and creates an adversarial environment in the relationship. It sends a strong signal that reason, co operation and collaboration will take a back seat to strict legal relationships, thus reducing trust between the parties. In other words, it is a strong potential source of conflict poison, making claims and conflict much more likely. As stated by Ramy Zaghloul at a recent Project Management Institute Seminar, "An additional but less visible cost of shifting risk to the contractor through disclaimer clauses [includes] …above all, more adversarial owner-contractor relationships."[6]

Risk should be fairly and reasonably allocated in the contract to ensure that both the owner and the contractor are motivated to the desired

---

[5] R. Zaghloul, F. Hartman, "Construction Contracts and Risk Allocation, Proceedings of the Project Management Institute Annual Seminar & Symposium" (October 2002).

[6] *Ibid.*

end — an on-time and on-budget project. The parties should do this by addressing these two areas:

1. *What are the Client's Responsibilities?*: The contract should clearly specify the responsibility of the client. Such items may include landscaping, building permits, fees or levies, architectural plans, letters of credit, *etc*. It should also state the amount and level of involvement in decision making the client needs or wants, along with anything else that is required of the client and in what situations.
2. *What are the Contractor's Responsibilities?*: The contract should clearly specify the responsibility of the contractor. Such items, for example, may include on site security, material shrinkage, hours of operation, part time labour, scope of work, quality of work, detailed product list, workers' compensation, liability insurance needed, and many more.

(vi) Other Key Contract Provisions

(A) Bid and Performance Bonds

Will the project be bonded? Who will provide these bonds? Who will pay the cost for these bonds? How long do these bonds need to be in effect?

(B) Change Orders

All items being done outside the scope of work will be billed as extras. Who needs to approve them? How will they be approved? When will payment be received for these additional costs? What is the extra cost to the contractor for supervising these changes?

(C) Payment Stream And Invoicing

When will payments be made? How often can a payment be requested? How long will it take to receive payment? Who will approve payments? This provision should include a very detailed payment system that is clear to all parties.

(D) Project Insurance

Who is responsible for insurance? What type of insurance is needed for the project? Who is responsible for the cost of insurance?

(E) Communication

How will the parties communicate (phone, fax, email, on site meetings, *etc.*)? How often will the parties communicate? Who is responsible to each party for this communication? How much authority do they have?

(F) Warranties

What is under warranty? What type of warranty is it? How long are these items under warranty for? Who provides the warranties (suppliers, manufacturers, trades, subtrades, *etc.*)? When warranty work is needed, who does the client talk to? How long will it take for warranty items to be rectified? Is any money being held back for this work? How long will this money be help back for?

(G) Deficiencies

Who is notified? Who is responsible? What is the timeframe for deficiencies to be remedied? What are the penalties, if any?

(H) Commencement and Completion Dates

When will the project begin? When will the project be completed? How will completion be defined and determined? Are there any bonuses for early completion? What are they and how are they determined? Are there any penalties for late completion? What are they and how will they be determined?

(I) Delays or Disclaimer Clauses

Ensure the contract defines the types of extensions allowable for delays without penalties, such as an act of God or *force majeur*, inclement weather making it impossible or impractical to proceed with the construction, labour disruption, material shortages, destruction of the dwelling by arson, mischief, or other third party intervention, additional work and/or changes to the original contract which result in extra time required to complete. This should be included as part of the risk allocation process described above.

(J) Value Engineering

Should the contractor be looking for ways to save the client money? If so, who gets the benefit of the savings? Clearly, if the contractor gets no benefit from saving time or money, he or she is not likely to even consider it. On the other hand, saving money by cutting corners or lowering quality will not be acceptable to the client. Define value engineering and how approvals will work and savings shared. Typical agreements for value engineering are:

- A minimum of $20,000[7] must be saved to even warrant consideration;
- No time to be lost to the changes;
- The same or better quality of product;
- Approval from the client must be in writing;
- Parties share net savings on a 50/50 basis.

II. SUMMARY

This chapter has provided an overview on the process of building trust and maintaining relationships before the project starts. It has covered the investigative work necessary in order to ensure a smooth transition from the bid phase through the contract phase, right up to the day the parties break ground.

In the next chapter we will focus on the construction process itself, and how the parties can stay focused on resolution if and when the conflict poison starts.

III. CASE STUDIES

(a) Case #1: Bridge Over Untroubled Waters

The following case illustrates how parties can use good preparation and preventative approaches to structure the project, and eliminate conflict down the road.

A major project was put to bid by government for the refurbishment and resurfacing of a bridge in Eastern Canada. The highway that the bridge was part of was one of the busiest sections of highway in the area, and length of project, completion date and road closures were critical issues for the government. In addition, there was another bridge of similar size within 20 miles that had also been refurbished and resurfaced recently. On that project, the work took 2.5 years to complete (one year past the contract deadline), and resulted in the government and the contractor suing each other for over $1.5 million. For this reason, the

---

[7] This figure is somewhat dependent on the size of the project, and the priorities of the owner.

government was highly sensitive to not repeating the problems on this project, especially since it was bound by tendering rules and had to choose the lowest bidder who met the requirements of the contract.

For this reason, the government chose to Partner the project. After the contractor was chosen (a different contractor than the one on the previous bridge project) and the Partnering date selected, the Partnering facilitators spoke to everyone during the pre-alignment process. At these meetings, everyone on the government side of the project, from the ministry representatives to the design consultants to the owner's representative, told the facilitators that even though the general contractor (GC) had bid the project to be completed within one year, the timeframe was impossible; the GC had bid it this way only to win the bid, and they all knew that the project could not be done in a year. They all believed that the GC had lied, and would find excuses for going over the deadline as it approached.

The parties met for two full days in the Partnering workshop. During this time, the parties worked on a charter, discussed the overall goals and objectives, and discussed specific challenges on the project (including the tight timeframes, *etc.*). During none of this work, however, did any of the owner's representatives or consultants publicly address the fact that they thought that completion in a year was impossible, or that the GC had bid the job knowing it could not complete it in a year.

On the afternoon of the second day, the group was discussing how issues could be resolved effectively. The owner's representative (OR) asked the GC's project manager (PM) for the name of their Site Supervisor (the Site Supervisor is the person directly responsible for all activity on the site at any given time). The following exchange took place:

OR: What's the name of your Site Supervisor?

PM: For what trade?

OR: No, sorry, not the trades, the Site Supervisor.

PM: I know, but for what trade?

OR: No, no, you're not understanding me. I want to know the name of the Site Supervisor for the project.

PM: I know, but it depends on which trade.

OR: No, it doesn't, it... (getting frustrated). Look, I just want to know who is in charge of the site.

PM: I keep saying, it depends on the trade, we have carpentry, electrical, concrete...

OR: (Cutting him off) No! The Site Supervisor! Who is it? It's a simple question!

PM: And I keep telling you, it depends! We have five Site Supervisors for different trades! We are going to be running a 24/7 operation with five different Site Supervisors!

OR: (Standing up and pointing) You can't run a job site like that!!

PM: (Standing up) We can and we will! <u>That</u> is how we're going to build this in 12 months!

At this point, the facilitators intervened, led a discussion on how the site was going to be run, who the points of contact were, how it would all fit with the Issue Resolution Ladder. At the end, the owner's representative quietly said, "Well, I don't know how successful this will be, but at least I understand what's going on."

Two things should be noted from this story. First, this project, which was budgeted at over $20 million, was ready to break ground within 5 days, and the key players had no idea how the project was actually going to be run. This is nothing if not a recipe for disaster and conflict. Partnering, as a preventative conflict management process, uncovered this issue (along with numerous smaller ones), and helped the parties address them up front, before they developed into disputes. Second, it should be noted that the project finished one week earlier than the 12-month deadline, came in on budget, and had no significant conflict during the project. The government, needless to say, was more than pleased.

(b) Case #2: The Case of the Missing Contractor

The following case illustrates how sloppy practice and assumptions can add up to disaster.

This was a multi-party *Construction Lien Act* case that included a general contractor (GC), subcontractor (SC) and owner. This case shows how multiple problems left unresolved can compound until no resolution is possible.

The situation was fairly typical of a construction project. The electrical SC had bid a job to the GC for $140,000. He had commenced work, and began invoicing for the work. He was paid $95,000 by the GC, at which time the GC abandoned the site and did not come back.

When this happened, the SC approached the owner, whose representative requested the SC continue to work and continue to submit invoices made out to the GC, and that the owner would pay directly. The owner's representative continued to direct the SC to do work, some of it apparently beyond the scope of the original contract. No change orders were written up, and the SC continued to submit invoices for all work done, which ended up totalling $220,000 (up from the original bid price of $140,000). The SC had invoiced $220,000, been paid $95,000, and claimed the difference, $125,000.

There was one additional relevant fact. It turned out that the Ontario government, 5 years prior to the execution of this contract, had cancelled the GC's corporation. In other words, the entity that had signed the construction contracts did not exist. This was relevant because without a legitimate contracting entity, the original contract was either void and unenforceable, or the GC had become personally liable.

In the lawsuit, the parties took the following positions:

The SC took the position that he had contracted with the GC's company, which did not exist, so that contract was void. This was important, because if the GC contract was valid, the SC would be entitled to only 10 per cent of the invoiced amount, which was 10 per cent of $220,000, or $22,000 in total from the owner. If the GC contract was not valid, the SC could claim, in theory, the full $125,000. Second, the SC took the position that after the GC left, the owner's representative had asked the SC to stay, and had directed the SC's activities. He argued that this evidenced a contract directly between the SC and the owner, and the owner therefore owed the full amount. Finally, he argued that the owner had been unjustly enriched by the SC's work, for which he was not paid.

The owner stated that all the SC was owed was the original bid price of $140,000, as it had been a fixed price contract. In that case, given that he had invoiced a total of $220,000, the most that the SC was entitled to was the 10 per cent holdback, or $22,000. The owner disputed that the additional work was outside the scope of the original bid, and additionally stated that the SC did not have any change orders signed that would give him the right to request more payment. Finally, the owner said that the contract was solely between the SC and GC, as all invoices were made out to the GC, so the owner was not obligated in any way to the SC, and invited the SC to go after the GC for the balance.

Finally, there were some major credibility issues, with the SC claiming that the owner's representative had promised to pay the SC directly, which the owner denied, pointing to the fact that the invoices were to the GC. The SC felt lied to and tricked. The owner stated that the rates charged by the SC on work after the GC left were inflated to gouge the owner, and stated that invoices contained work that had not actually been done, though the owner had no objective proof of this.

By the time this reached mediation, it was well after the project had been completed. The owner was not held up by the lien claim, and the SC had moved on to other projects, so neither was strongly motivated to settle for external reasons. In addition, through the litigation process both parties had made serious allegations including fraud and bad faith actions. Both parties at this time were trying to teach the other party a lesson. In the opening stages, the venting took on a personal tone. There was little objective information to confirm many of the allegations. After 4 hours of exploring options, it was clear that money was the only issue, and neither would move off its position since they viewed the other party as unethical and dishonest. It went forward to trial.

(i)   Lessons from the Case

There are a number of lessons from this case:

1. **Who is the Contract with?:** Had the SC looked into the GC's corporation, he would have quickly found out that there was no corporation, and the whole issue of who the contracting entity was would have been dealt with up front. In addition, when the owner's representative began directing his work, he should have recognized that the contracting parties had changed, and this required acknowledgment in writing.
2. **Change Orders**: The SC should never have continued doing work that he believed was outside the scope of the original contract. Had the SC requested signed change orders at the time of the additional work, the issue would have surfaced before he did all the work, and significantly reduced the size of the claim. It also would have given him proof that the extras were just that, extras, and removed the owner's argument that all work was part of the original contract.
3. **When to Mediate**: This mediation happened far too late. The GC was long gone, but so was the owner's representative who could have shed some light on the whole process. In addition, if the owner needed the project completed, he would have been more motivated to resolve the claim. Had the SC still been on site, he would have been more motivated to reach a resolution and move on.

4

BEST ADR PRACTICES — DISPUTE RESOLUTION DURING THE CONSTRUCTION PROJECT

The single biggest challenge that parties have during construction is maintaining strong working relationships throughout the project. Construction projects are high stakes and high pressure for everyone involved; because of this conflict is rarely a question of "if" it occurs and more a question of "when" it occurs. Projects where working relationships are strong will confront and resolve disputes far more effectively than projects where the parties communicate little and trust less. The quality of the parties' relationships, in effect, will determine how successfully day-to-day issues are dealt with. The key areas for maintaining strong relationships and for minimizing the flow of conflict poison are:

1. Clear, open and proactive communication;
2. A commitment to resolving disputes effectively when they arise.

We will look at each of these areas and how to best implement them during the construction process.

I. COMMUNICATION

Clear and open communication is something we have talked about many times in this book, and is one of the strongest antidotes to conflict and conflict escalation. When talking about communication, it is easy to focus on the content of the communication itself, on *what* needs to be communicated on any given day. While this is important, focusing primarily on the content and not on an effective way of communicating is one of the biggest causes of failure on a project.

Effective communication in construction starts with a clear and consistent structure and process for communication throughout the project. Effective communication builds the paths and channels for communication, and then requires all parties to participate in those channels throughout the project. By creating the communication pathways and structures, effective communication begins to happen with less and less effort. It is primarily the job of the project manager to ensure that the communication channels are clear and unobstructed at all times. This means continually monitoring and guiding the communication process.

In terms of communication, construction projects are similar to a computer network centered around a mainframe computer. A mainframe computer network only operates if the network is "up" and running; in other words, all the computers in the network can access and talk to the mainframe (and each other) effectively. To ensure this is happening, the mainframe continually "shakes hands" with its terminals in order to ensure the path for data transfer (communication) is not becoming obstructed. If an obstruction occurs, the mainframe will immediately try and solve the problem, rather than wait until that pathway is needed at a critical time. This same communication "handshake" must take place as a way of monitoring the quality of the communications process on a construction project. The way the communications path is continually being checked in a mainframe computer is through diagnostics and preventative maintenance. Just like in the computer, a construction project must also run diagnostics and preventative maintenance throughout the project. This process on a construction project is done through various types of meetings.

(a) Daily Site Meetings

The first level of communication is the daily meeting, which typically takes place on site. These meetings can be formal or informal and will be focused on the most immediate issues the parties have in getting that day's work done. While attendance at the daily site meetings will vary depending on the nature of the work being done, these meetings should be required for everyone working on site that day.

Frequently, these meetings are used by the site supervisor to communicate about anything that will impact that day's work. In addition, the site supervisor gathers updates on progress and any problems or issues anticipated by anyone. This is the front line diagnostic and preventative maintenance for the project, the first opportunity to uncover problems or potential issues.

There is a strong tendency to avoid these meetings and have the site supervisor simply "make the rounds" to talk to each of the trades individually. This is often where the communication process begins to break down, for a number of reasons. First, it signals that communication is done on an *ad hoc* basis, and therefore is not that important. Second, it tells the parties their job is to mind their own business and get their work done, rather than to participate effectively in the whole process. In other words, it isolates each party in their own little world rather than requiring all parties to be an important part of the whole project. Finally, it places the entire communication responsibility on the site supervisor, as he or she is the only person who has talked directly to everyone on site. When the site supervisor becomes the sole hub and filter for all communication, it sets him or her up for being the scapegoat should anything go wrong, encouraging a process of blame and finger-pointing rather than problem solving.

Daily Site Meeting Summary

Meeting Frequency:	Daily
Who Attends:	• Site supervisor; • Architect/Consultant (as needed); • Subs currently working on site; • Anyone else with information/issues relating to the day's work.
Length:	15 – 30 minutes
Agenda:	Informal, can be handed out at the meeting. Typically includes: • Issues for the day; • Production delays; • Inspection requirements; • Delivery issues; • Potential obstacles; • Issues looking ahead; • Challenges/concerns for anyone; • Project status; • Wrap-up.
Documentation:	Site manager keeps brief notes of decisions made, responsibility, and timeframes.

Site meetings should be held daily, chaired by the site supervisor, and documented in simple clear notes covering what was discussed (in summary fashion), what decisions were made, who is responsible, and by when. A sample daily site meeting record is below:

SAMPLE DAILY SITE MEETING MINUTES
Daily site meeting held on (date and time)
Location (site construction trailer)

In attendance

R. Smith, Site Supervisor

J. Jones, Rough Framers

L. Collings, Plumbers

M. Michaels, HVAC Company for rough-in

K. Toms, Electrician

Discussions	Responsibility	Follow Up
Framers need second floor materials tomorrow	Site supervisor	Order materials
Plumbers have rough-in inspection today at 1300 hrs. Everything looks good to proceed with water test.	Plumber	Update tomorrow
HVAC rough-ins to be complete in two days. Waiting for gas company to inspect gas lines.	HVAC company	Update tomorrow
Bricks need to be moved to allow for lumber delivery.	Site supervisor	Update tomorrow
Temporary power and lights required for units 4 and 5 to allow for night shift work commencing this weekend.	Electrician	Update tomorrow
Reminder about safety issues.	Site supervisor	On site inspections
Next meeting at 7:00 a.m. tomorrow (date).		

(b) Weekly Site Meetings

The weekly site meetings are the backbone of good communication on the project. Weekly site meetings perform a number of critical functions, on both the project level as well as on the relationship level. Without effective weekly site meetings, it is only a matter of time before major conflict occurs. The basics for effective weekly site meetings are:

(i) Attendance

The weekly meetings should always be attended by the appropriate representatives of the contractor, the owner, and other important stakeholders, such as key subcontractors or suppliers. For the contractor, the site supervisor, site manager and/or project manager must attend and run the meeting. Note that the size of the project will decide who is appropriate; on smaller projects, the general contractor may regularly attend, while on major projects the site manager or project manager may be the senior representative. For the owner, the owner's representative must attend, along with any owner's consultants, as appropriate. Key subcontractors currently or about to be on site should also be present. If the major stakeholders or contractor require other parties to attend this meeting they should give written notice to those parties prior to the meeting requesting their attendance. If any other parties involved in the daily meetings wish to attend the weekly meetings for a specific reason, this request should be put in writing prior to the weekly meeting. This format will ensure proper time allocation for setting the agenda and the meeting itself.

(ii) Meeting Agendas

The structure of the meetings is critical — they must be detailed enough to be effective, and short enough that all parties feel that their time is well spent. The agenda at these meetings should be set in advance, and should clearly lay out the topics or matters of business that need to be discussed. The agenda should indicate the actions needed, such as "information only", or "decision needed". Every item discussed should have an action associated with it and include a follow up date, action plan and designation of responsibility, or subsequently be documented as complete. The agenda should have fixed items to ensure continuity week to week. For example, each meeting should begin with a follow up from the last meeting's action steps before discussing any new business. This will ensure important issues will not fall between the cracks and get lost.

(iii) Weekly Site Meeting Minutes

Weekly site meetings must be documented, and these minutes should be distributed widely to ensure everyone is kept informed. The minutes form a continuous record of the project that keep everyone in touch with the status of the project, and alert parties to issues as early as possible. The minutes also document what was decided, what follow up is required, and who is responsible for the follow up by when.

Suppose, for example, there is a discussion at the weekly site meeting with regards to the installation of the elevator in the building. The elevator company expresses concern that the delivery of the elevator may be delayed by two weeks by their supplier. This impacts the contract with the kitchen company, because that contract clearly states that the cabinets will be delivered to the second floor via the elevator, not the crane. The project manager needs to address this issue.

First of all, the parties impacted by this possible delay who must be notified are:

- The financial institution, which must be informed since a delay in overall completion will impact the funding and cash flow.
- The owner, who must be informed of the possible late closing, and the effects on his or her cash flow.
- The purchasers, who need to be notified of the possible late closings so they can find somewhere to live and store their furniture for two weeks.
- The project manager, who must change the scheduling for the remainder of the project or ask for more money to pay overtime and catch up.
- The kitchen company, which must find a place to store the kitchen cabinets until they can be delivered and reschedule the necessary labour to do the installation.

As many of the above parties need to be notified to attend this weekly meeting as possible. At the meeting where the discussions have taken place on the impacts of this delay, next steps and decisions need to be looked at. Let us look at how the minutes and action plan of this meeting should be documented.

Minutes of meeting held on (date and time)

Item # 5 — Delivery of elevator delayed by 2 weeks

Action Step	Responsibility	Done by/Follow up (date)
Confirm actual delivery date	Elevator company	24 hours
Notify financial institution	Owner	24 hours
Inform customers of late closing	Owner	24 hours
Change production schedule	Project manager	48 hours
Storage of kitchen cabinets	Kitchen supplier	24 hours

This shows how issues that are raised should be documented and focused into actions that minimize the impact of the problem. By documenting the discussion as above, it keeps the discussion focused on problem solving, and eliminates much of the "he said/she said" that emerges when parties are communicating poorly.

Weekly Site Meeting Summary

Meeting Frequency	Weekly on-site
Who Attends:	• Site Manager, Project Manager, Contractor; • Owner's Representative; • Architect/Consultants (as needed); • Subs currently working on site; • Anyone else with information/issues relating to the day's work.
Length:	30 - 90 minutes, as needed
Agenda:	Set ahead of the meeting, with fixed agenda items such as: • Previous action items; • Status of current issues; • Project status; • Current business (submitted ahead of time, if possible); • Future issues/looking ahead; • Next steps
Documentation:	Formal minutes kept, including: Next steps, responsibility, deadline, etc. Minutes distributed within 24-36 hours of meeting.

SAMPLE WEEKLY SITE MEETING MINUTES

Weekly site meeting held on (date and time)

Location (site construction trailer)

In attendance

R. Smith, Site Supervisor

J. Williams, Director, New Construction

J. Roberts, V.P. Marketing and Sales

S. Ralf, Project Manager

K. Reve, Architect

T. Johnson, Site Manager, Steel Erection

D. Webster, Sales Manager, Prefab Steel

Discussions	Responsibility	Follow up
No outstanding items from last meeting		
New Business		
Mr. Johnson is concerned that steel delivered to site does not fit on the foundation walls, causing his crews and the project to be delayed. He wants to know what is going on.		
Mr. Ralf said he has ordered materials from the blueprints and it should fit properly on the foundation. He will recheck his takeoff that he has given to Prefab Steel.	Mr. Ralf & Mr. Johnson	Today
Mr. Reve will do a site inspection to compare as built construction with his blue prints.	Mr. Reve	Today
J. Roberts needs to know how long this will delay the project so he can inform his sales staff of new closing dates.	Mr. Smith	Tomorrow
Mr. Webster has built and delivered the steel as per the purchase order given to him by Mr. Ralf. He will rush the amended order once determined by Mr. Ralf in order to minimize delays on the project.	Mr. Ralf	Tomorrow
Mr. Williams wants an emergency meeting held tomorrow with answers and an action plan for this problem so he can report to head office.	All Parties	Tomorrow
Mr. Smith will inform other trades of possible delays and prepare new production schedule.	Mr. Smith	Tomorrow

Emergency Meeting Tomorrow at 9:00 AM (at head office)

(iv) Senior Management Meetings

The final meeting level takes place on larger projects only, and is designed to build and maintain strong relationships at the highest level on the project. In many cases, once a project is bid and ground is broken, senior people at the major stakeholders will not talk or see each other again until the end of the project, unless a significant issue arises. If and when a major issue arises, they have virtually no knowledge of or relationship with their counterparts. The president of the company building the office tower might never have met the owner, president, or even vice-president of the national construction company chosen to build the tower. Suddenly, with a major issue to deal with, they need to sit down and work on a solution. Not surprisingly, the first call both parties make is to their counsel, rather than their counterpart.

Senior management meetings should take place regularly on the project as a way of establishing communications and relationships at the highest level of each major stakeholder. The focus of these meetings is less on the project itself and the daily or weekly status (although that is an important topic of discussion), and more on how the project is proceeding in general and how happy all parties are on the project. This is a significant point, in that senior managers will have staff who handle the "what" of the project, *i.e.*, what is built, what still needs to be done, where the budget stands, *etc*. This is crucial information, of course. In addition, however, senior management should also be very concerned with the "how", *i.e.*, how effectively parties are working together, how timely the communication is, how happy everyone is, and how the senior management can help.

By having this discussion regularly on the project, should a significant issue arise that requires senior management to intervene, a strong relationship and good communication already exists and can help immeasurably in finding a resolution to the problem. Without this level of meeting taking place, the courts are the most likely next stop for the parties.

Senior management meetings should take place every one to three months, with a simple agenda that focuses on the status of the project from a high level, both parties' perceptions of how well it is going along with concrete examples, and what could be done to improve the working relationship for both parties.

Senior Management Meeting Summary

Meeting Frequency:	Monthly to Quarterly
Who Attends:	• Owner/President/VP of Contractor • Site Manager/Project Manager • Owner/President of Owner • Owner's representative
Length:	60 - 90 minutes
Agenda:	Typically includes: • Project Status, high level • Perceptions of how well project is going, and why • What could improve the process for both parties • Wrap-up
Documentation:	Simple minutes kept documenting what was discussed and any decisions taken.

(v) Summary — Communication

Communication is the lifeblood of the construction process, and effective meetings are absolutely necessary to ensure that timely communication happens. The single biggest mistake that is made on construction projects is the assumption that communication will take place of its own accord, that communication simply will happen. It will not. Human nature is such that when difficult issues arise, communication is the first thing to disappear. When serious issues arise on the project, people have a tendency to duck and run for cover rather than be the bearer of bad news. Since ancient times, the bearer of bad news, the "messenger", tends to get punished for bringing the unwelcome message, and this dynamic, this belief, causes many parties to put their heads in the sand and hope for the best. Allowing this to happen invites trouble.

Effective communication can be achieved, however, by building and maintaining simple, effective meeting structures. As the saying goes, "If you build it, they will come". If you build good meeting structures, simple agendas and documentation along with clear follow up, good communication will follow.

A final caution on communication — do not let success ruin it. One of the strongest tendencies is to abandon the discipline of regular meetings when there appears to be few issues to deal with. Good communication is a required infrastructure that needs to be there when things get difficult, so the meetings should be maintained whether some of the stakeholders feel they are needed or not. Holding short, effective meetings is a small investment of time, and by doing so you will be inoculating the project against major problems in the future.

II. ISSUE RESOLUTION PROCESS

Effective issue resolution is critical to preventing and minimizing conflict, as well as eliminating the poison to the relationship that conflict sometimes causes. In this section, when we talk about an issue resolution process, we are referring to a voluntary, agreed approach to issue resolution that parties will use prior to accessing more formal processes, such as the contract or the law. In reference to Chapter 2, Figure 2.1, we are specifically talking about an early intervention on the interest-based stair, before parties move up to any rights-based processes. Once the contract or the law become the primary dispute resolution process, the conflict poison will tend to flow out of control.

There are two reasons that issues on the project turn into disputes:

1. *They take too long to address*: This can happen for a number of reasons. The site supervisor may decide the issue is unimportant or unreasonable, and ignore it in the hope it will simply go away. There may need to be input from technical experts, and they are unavailable or slow in responding. It may take too long to pass it up to a higher level. There may be personality clashes between the people handling the problem, resulting in avoidance by one or more parties. In all of these cases, the delays cause the problem to get worse, for the costs to mount higher and higher, and give every party more reason to want to blame the other for the problems.
2. *The wrong people are trying to solve the problems*: It may seem obvious to say that the right people need to be working on the problem; what is not immediately obvious when issues arise is who the right people are. Generally speaking, the people closest to the problem are best suited to finding a good solution, and they should be the first to address problems as they arise. That said, some solutions would be best handled by escalating them to the highest levels as quickly as possible. The challenge is finding a way to do both of these effectively.

There are a few best practices that can help address both of these reasons that cause problems to escalate into disputes.

(a) Issue Resolution Ladder (IRL)

One of the most effective issue resolution processes on a construction project is the Issue Resolution Ladder (IRL).[1] The IRL is a tool that parties design and agree to before the start of the project. Essentially, the IRL is a stepped negotiation process that identifies (1) who will be responsible for addressing and resolving issues at each level of each organization on the project, and (2) how long each level has to resolve a dispute before it automatically escalates to the next level for help.

For example, on most projects, the first person to address issues for the general contractor is the site supervisor, who is in charge of the site on a day-to-day basis. For the owner, this is typically the owner's consultant or representative, the person who inspects the work as *per* the plans everyone is working from. Above the site supervisor is the project manager, and above the owner's consultant is the architect or possibly the engineer. Above the project manager is the owner of the construction company, and above the architect is the actual owner. Now, if we put these three levels on a diagram, it looks like this:

Figure 4.1 — Stepped Negotiation Diagram

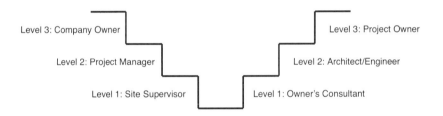

What Figure 4.1 illustrates is that when an issue arises, it is typically addressed by Level 1, the site supervisor and the owner's consultant. In many cases, it is resolved there. In cases where conflict escalates, what happens is this. An issue arises, and the site supervisor asks the owner's consultant if he or she can make a minor change to the process or plans. The owner's consultant, charged with ensuring the plans are followed, refuses. This answer does not make sense and does not work for the site supervisor, so after repeated attempts and arguments with the consultant the supervisor simply calls the architect or the engineer and pleads the case with him or her. When the owner's representative finds out about

---

[1] In Chapter 3, Partnering was discussed. One of the most useful outputs from Partnering is an Issue Resolution Ladder.

this, he or she feels that the supervisor has done an "end run" around him or her, and is not disposed to help that supervisor (now or in the future). This is the start of the conflict poison. Further, reports filter up the ladder to the owner that the contractor's people are difficult and untrustworthy, setting the stage for more problems. The architect then feels some pressure to "back up" the consultant's decision, rather than solve the problem on its own merits. The conflict is now escalating out of control.

The IRL tries to change this dynamic, and it does this by formally identifying each person at each level for each important organization and placing that person on a similar ladder above. It then defines a specific period of time each "level" has to hold on to and resolve an issue before others get involved. After that period of time, the issue automatically escalates to the next level for both parties. A typical IRL, with time frames, is this:

Figure 4.2

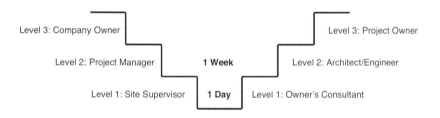

Here is an example. The plans call for mechanical equipment of a particular specification. The supplier informs the mechanical contractor that the required model is no longer available, and suggests a replacement that is similar. The mechanical contractor asks the site supervisor for approval, who turns to the owner's consultant for an answer. The owner's consultant refuses to approve it, saying the supplier has to provide what was in the contract. They are at a stalemate. The mechanical contractor must order the equipment within the next few days, or it cannot be delivered on time.

Without an IRL, it would not be unusual for the issue to be argued about for days at the front line, until the site supervisor goes to the project manager for help. Depending on the quality of the relationships, the project manager may (in the best case) go directly to the architect or engineer, leaving the owner's consultant upset that he or she has been bypassed. In the worst case, the project manager goes to the company owner, who then goes to the project owner, who then needs to gather information from the architect (who is upset that he or she was not approached first). All of this can take a number of days, if not weeks.

With an IRL, the site supervisor and the owner's consultant have 24 hours to agree to a solution, or a process to get a solution. After that, it is automatically escalated by both sides to their bosses, the project manager and the architect. They then have up to a week to resolve it, or it automatically goes up to both owners. Since this process is clear and agreed to up front, no one will feel that an "end run" has unfairly occurred.

The IRL basically does two things.

1. It maintains a downward pressure for issues to be resolved at the front line. No one likes to go to his or her boss and say they cannot resolve something. They are professionals who take pride in their ability to solve problems. Because issues are required to escalate quickly, both parties pay more attention to getting a resolution before they automatically escalate.
2. It escalates serious issue quickly. Because Level 1 people address issues quickly, they will also quickly recognize when an issue is serious, and will get it up to the right level of authority much faster, minimizing damages caused by delay.

What is important to note about the IRL is that it functions solely as a backstop to catch and address issues that are not being handled effectively and quickly. It does not mean that all issues must be handled in the timeframes on the IRL, since many require a longer timeframe in order to get expert advice, testing, *etc*. It does mean, however, that parties have to have an agreed way forward within those timeframes. In other words, the IRL is what any party can turn to when it does not know what else to do, when there is not a way forward that the party agrees with.

Below is an actual IRL from a medium-sized sewer upgrade project. The names have been changed, but the structure is identical to the IRL used on the project.

Figure 4.3: Issue Resolution Ladder Developed by the Parties

Time Frame Level 1-4	Ministry of Environment	Town of New Bedford	Filmore Associates Architect	Oak Contracting Corp.	Northern Excavation
1 hour (Level 1)	N/A	N/A	J. Babb, Inspector	S. Tomkins, K. Calvin, Site Supers	N/A
24 hours (Level 2)	N. Smith, Inspector	G. Root, Manager of Development	C. Delmer, Site Architect	V. Gregson, Project Manager	A. Mario, Equipment Supervisor
72 hours (Level 3)	D. Manna, Manager of Inspections	Z. Serling, Town Councillor	J. Coombs, Design Architect	S. Phillips, VP Construction	F. Beeton, Owner
(Level 4)	N. Wills, ADM	Town Council	S. Filmore, Partner	R. Van Demm, Owner	F. Beeton, Owner

On the above IRL, note a few things. First, the primary interface is between Oak Contracting as the general contractor, and Filmore Associates as the architect and owner's representative. In addition, however, the owner (New Bedford), the Ministry of the Environment, and Northern Excavation are also on the IRL, since they will be key players in both the work itself and any dispute that arises. Second, note that there is no timeframe on the final or top level. Once a dispute is with the various owners of the companies, the buck stops there. If they want to take a month, they will. The key is to get it to the highest level quickly to give everyone the opportunity to resolve the dispute before significant delays and damages have occurred.

The IRL is a tool used to define exactly who is in control of issues, and how those issues will escalate. On successful projects, the IRL is rarely invoked, since parties have a strong enough relationship to address issues effectively as they arise. By developing and committing to the IRL, however, it gives parties a clear path and process for both resolving issues at the lowest level possible, as well as escalating the issues in a safe and agreed way to the proper level for resolution.

(b) Using a Standing Neutral

Another approach to dispute resolution during the project is the appointment of a standing neutral on the project. We have looked at the different possibilities for the standing neutral in Chapter 3. Here, we will look at the effective uses of the standing neutral during the project. There are a number of reasons for involving the standing neutral during the project, and they are similar to the use of the IRL. As long as parties are engaged and working together effectively, there is no need for the involvement of the standing neutral. Parties should engage the standing neutral to help with an issue when:

- Communication or engagement in working on the problem begins to break down, and hard feelings or an "us vs. them" mentality begins to develop;
- Delays are threatened;
- Additional costs have or will result from the issue;
- Relationships are strained and not getting better;
- The parties are more interested in blaming one another than finding a solution.

Upon getting involved, the standing neutral will facilitate communication between the parties, help clarify positions, focus the parties on the overall goals and interests, and assist in assessing the parties' approach to resolving the disputes. Once the issues are dealt with and agreed on, the focus can return to completing the project; through this process, the relationships will likely be strengthened.

(c) Other Dispute Resolution Issues During the Project

(i) Change Orders or Extras

Being paid for work done beyond the contractual obligations on a construction project is one of the largest areas for dispute, and frequently results in severed working relationships and litigation. The problem itself is simple: by the time an invoice for extra work is submitted to the general contractor (frequently without the proper documentation), the job is usually completed. By that time, the extra work is forgotten and the price or value hotly disputed. This should never happen, for any reason.

When any work outside of the contract is requested, there should be a formal procedure in place that specifically identifies the work to be done, the timing, and the cost of the work. No extra work should commence without approval, in writing, by the parties who have the authority to approve this work. Even if there is an urgent request for changes to the contracted work, it should not proceed until approval in writing is granted.

One method of insuring that extras done are not forgotten or disputed is to insist on purchase orders (PO). The PO should include the date the extra work was requested, a description of this work, the cost of doing this work and the signatures approving the extra work and cost.

This is an area where the quality of the relationships can actually lead to trouble. When a good working relationship has been built over time between the general contractor and subcontractors, it is likely assumed that if the general contractor informally requests additional work, the sub will be paid for that work regardless. This is a recipe for destroying the relationship between the parties. While the general contractor and subtrades may work together on many different sites, always remember that the party paying the bills, *i.e.*, the owner, has no such relationship with anybody. The owner is entitled to understand and approve all costs associated with extra work being done, and when invoices are presented long after the fact they will likely go unpaid.

(ii) Value Engineering Process

In some situations during the building of the project, the general contractor and its subtrades may offer cost saving options to the owner or owner's engineer. The type of materials being used, new technologies now available, or the layout of the construction project could be discussed. This is a value-added advantage for the benefit of the customer and should be encouraged by way of remuneration to the parties involved in finding these savings. This is a win/win situation for all parties involved and allows for the input and expertise of each party to be used effectively in the project.

In the tender documents there should be a clause that allows for a value engineering process that encourages better methods of work completion without jeopardizing the quality of work performed or the time it takes to complete. It is important, however, that the parameters and guidelines for any value engineering proposal be established up front. Typically, these guidelines should include:

- *The minimum savings for consideration*: It costs time and money to make changes to the plans, and savings of $1,000, for example, actually cost more than they save. In many cases, minimum savings are dependent on the size of the project, but are often in the $15,000 to $40,000 range.

- *No reduction in quality*: An absolute guideline is that the quality of the proposed changes must at minimum be equal to, and ideally enhance, the quality level of the current design.
- *Minimal or no delay in implementing the proposal*: Time is money, and if a change in technology or design saves the owner $100,000, but delays the project three months, odds are the cost outweighs the savings.
- *An agreed split of the savings*: Clearly, an incentive must be in place for the contractor to even consider value engineering ideas. Typically, the contractor proposing the value engineering savings receives between 10 per cent and 50 per cent of the savings generated.

It can be in everyone's interests to find better and cheaper ways to build the project. To prevent disputes and hard feelings, have clear value engineering processes in place for the project.

III. SUMMARY

Resolving disputes during the building phase of a project is a dynamic and high-pressure process. By establishing best practices around communication, meetings and issue resolution, many problems can be caught and addressed as they occur. This will result in early resolution, low cost solutions, and stronger relationships in the future.

IV. CASE STUDIES

(a) Case #1: Smokin' the Big Pipe

The following case illustrates how parties can address difficult issues during the project, using effective ADR tools.

A municipality bid and awarded a contract for laying a new sewer main running approximately 34 kilometres in northern Ontario. The project had been Partnered, and over the 2-day partnering session, a few key concerns came up. First, the municipality reiterated over and over that it did not want to see any extras. It had had bad experiences in the past being presented invoices for extras during and after the project, and it wanted to avoid this completely. The contractor agreed.

Second, the municipality had retained an outside firm to inspect the work and ensure it was done to the quality level in the contract. The contractor had requested regular meetings with the inspectors to ensure that everyone was on the same page regarding quality as the project progressed. At the Partnering session, it was agreed that there would be strong open communication about any issues that arose to prevent any type of escalation.

The project got underway, and during the first two weeks, after the contractor had laid the first few kilometres of pipe, the inspector presented evidence to the site supervisor that they were using substandard fill on the project, and informed the owner at the same time. Further, the inspector had been on site frequently, had documented the substandard fill, and had attended two meetings on site without saying a word about the problem, simply allowing the contractor to continue doing the work with substandard fill. The site supervisor reacted angrily at the inspector, stating that the inspector had acted in bad faith by allowing them to continue doing the work without raising the problem. The inspector responded that he was not hired to "help the contractor, but just to report to the owner". It appeared that the project was off to a poisoned start. The site supervisor immediately invoked the Issue Resolution Ladder and got the owner and the contractor involved.

As part of the Partnering process, there had been "Partnering champions" appointed from each party to address issues if they arose. When this particular issue arose, the parties (somewhat reluctantly) asked the champions to convene a meeting later that day to decide what to do. At this first meeting, the contractor agreed to stop work for a day and dig test holes in the completed work to find out where there was substandard fill, and to share that information. This revealed that substandard fill had been used in a few places, although not as extensively as suggested by the inspector. At the second meeting of the parties' representatives (the champions), a number of solutions were brainstormed. First, all parties committed to fixing the problem to everyone's satisfaction as the goal, and second, parties committed to using the inspection process in the future to assist the contractor achieve the quality required, and not play "gotcha!". During this second meeting, the municipal representative mentioned in passing that a long-term need was a water main upgrade. Upon hearing this, the contractor stated that he had thought about suggesting to the municipality that it consider laying a water main in the same trench, as it would cost a fraction to lay the main in the same trench than it would re-trench in the future, but had not raised it due to the "no extras" position the municipality had taken. After looking at all the costs, both for the remediation and the new water main, it was agreed that the contractor would reopen all of the work done to date, replace any substandard fill, and lay a new water main the entire 34 kilometres at an agreed extra cost.

(i) Lessons from the Case

There were a few lessons the parties learned from this situation.

1. **Issue Resolution Ladder**: By having the ladder in place, it made it easy to get the owner and contractor involved without triggering a strong adversarial response from each other.
2. **Communication Structures**: It was extremely helpful to have communication structures built into the relationship to rely on when conflict arose. By having the Partnering champions from each party to call on, the parties were able to focus on looking for solutions, rather than looking to lay blame.
3. **Focus on Business Solutions, not Legal Problems**: With the help of the champions, the focus shifted to solving problems, not on continuing a game of trying to "catch" each other. In other words, the business issue of quality took precedence over the legal issue of blame.

(b) Case #2: Now You See Me, Now You Don't!

This project was the construction of a new high-end custom home with a fixed contract price of $1,400,000 and 11,000 square feet in size. The plans that were used to bid on this project were very detailed including structural, landscaping, elevation, mechanical, interior design, and water features. There was an existing structure on the property that needed to be demolished as well.

As part of the contract, it was the owner's architect who was responsible for obtaining all permits for construction. A start date was set, and the owner committed to having everything ready to go. The builder arranged for and scheduled the proper trades for the demolition and footing excavation for this commencement date. The day of the ground breaking the equipment showed up in order to do the demolition of the existing house and to dig the footings of the new house. The surveyor and the garbage bins required to remove the debris off site from the demolition were also in attendance. The carpenters followed in order to form the footings. The forming crew, cement trucks, and pump truck along with the Town inspector were scheduled to attend the site that afternoon. When the builder requested the permits, he was told that the architect was at the city picking up the permits.

Upon arriving, the architect had only the building permit, but had been declined for a demolition permit, as the house was deemed a historical building and would have to be moved off site and preserved. The builder was irate, but luckily the property was large, and the new house was located well behind the existing structure; with some creative approaches, they could start on the new structure while the status of the existing structure was resolved. The builder pressed ahead, and the project started, albeit a bit late.

Everything went smoothly for the footing and foundation pour. The first milestones were hit, and invoices were submitted and paid. Over the next two months, the owners stopped attending the weekly site meetings. The next phase was nearing completion when the bank monitor arrived to inspect the work. When the builder asked when the next draw would be paid, the monitor said that the draw for this work had already been paid, and no further advance would be made until the project was roofed and dry. The builder was surprised, and called the owners. Once again he was assured everything was fine, that they would talk at the next site meeting early in the week. Unfortunately, the owner did not show. The builder now called the owner and demanded to know what was going on. In addition, he demanded that the current invoice be paid, which totalled $250,000. The owner said he was travelling but would have his office forward a cheque for $40,000, the balance of which would come when he returned in 6 days. The builder agreed, and continued to work on the site.

At the end of 6 days, there was again no word from the owner. Worse, the cheque for $40,000 bounced. Again the builder called, angry, and threatened to stop work on the project. The owner apologized profusely, promised to replace the cheque, and to meet with the builder at the end of the week. The builder refused, and demanded a meeting the following day. The owner reluctantly agreed.

When the builder arrived for the meeting, he was greeted with the owner and the owner's lawyer, who told him that the owner had significant quality concerns and would not pay any further invoices until the quality issues were addressed. The builder suspected that this was just a delaying tactic, refused to proceed until all outstanding invoices were paid, and left. He immediately registered a construction lien on the property, and pulled his team.

Subsequently, it was discovered that the missing funds had been used to build a second custom home not far away, and both projects had run out of money. The bank had stopped advancing funds, and threatened to repossess both homes. The project sat idle for over 18 months, and the owners could not be found.

(i) Lessons from the Case

1. **Do Your Homework**: The builder should have monitored the funding process, known who the bank monitor was, and been in closer touch with him or her. Had the builder known more about the draw process,

he would have known when the draws were made and could have found out much earlier that there was a problem.
2. **Site Meetings**: Had site meetings been a higher priority, and had the builder made it clear that the owner was required to attend, these problems would have arisen much sooner and the builder would have lost significantly less money. By allowing poor communication and poor performance by the owner, it enabled the delays to mount, and with it the builder's losses.
3. **Communication**: It is not uncommon for the architect to be responsible for obtaining permits, but the builder should never have been surprised that the permits had not been granted. Had there been better communication early on, the builder would have either established clearer, more effective communication processes, or would have discovered the lack of co-operation and commitment earlier on.

5

BEST ADR PRACTICES — DEALING WITH DISPUTES AT THE END OF THE PROJECT

The final step in looking at best practices in construction ADR is what to do at the end of a project when a claim is outstanding. The work is complete, the project finished, but the books cannot be closed because of outstanding claims.

Construction is a difficult business, it is complex and demanding, and at the end of a year or two of hard work, the last thing many parties have the energy for is resolving a significant claim. It becomes easy, in a sense, to turn that dispute over to the legal system and let it run its course. At the time, it may actually appear to be the simplest thing to do.

A significant fact that is often overlooked when it comes to litigation is that close to 97 per cent of all lawsuits settle before a verdict is rendered at trial. This means that the odds are very high that any given claim will settle before trial, and it is really just a question of when and how the settlement will occur. It is odd, then, that mediation, negotiation, arbitration and other forms of settlement are considered "alternatives", is it not? In reality, litigation is the alternative to resolution and settlement, and should be treated as the alternative of last resort, not the alternative of first resort.

There is a wide range of choices that parties have for working on and achieving a settlement to outstanding claims. In the realm of ADR, the choices are almost endless; at last count, there were over 50 different processes to choose from. That said, these choices can be boiled down to one of two specific types of processes that we discussed in Chapter 2, specifically Interest-based or Rights-based processes, or some blend of the two. All of these processes have strengths and weaknesses, and parties are well served to choose the process that will best meet their real interests.

In this chapter, we will look at some of the most common and most used ADR processes, and how they can best be applied to construction claims.

I. MEDIATION

Mediation is a structured negotiation process in which an external, impartial third party assists two or more parties to reach an acceptable agreement to their dispute. Mediation is an interest-based process, and has many of the same characteristics of negotiation, namely:

- The mediation process is non-binding — resolution happens only if all parties agree to the settlement;
- The mediation process is voluntary, and the mediator has no formal authority or power over the parties. Parties are there solely to see if they can get their interests met, and can leave the process at any time. Even in Ontario's "mandatory mediation" program, what is mandatory is that the parties show up; after that, the parties may leave any time they wish;
- The mediation process is confidential and legally "without prejudice". This means that anything said or any offers made cannot be used in any way by any party in the future, should the matter not settle.[1] The obvious reason for having the mediation process off the record is to encourage frank and open discussions amongst the parties to effectively explore settlement.

Mediation can be triggered or accessed in different ways. Frequently, as discussed in Chapter 3, mediation is part of the dispute resolution clause of the construction contract. In other words, parties are mandated to attempt mediation because of the contract they have signed. In these cases, similar to a "mandatory mediation" program in the courts, parties are required to attempt mediation, but settlement will only occur if the parties voluntarily reach an agreement. Another way that is becoming more and more common is when parties jointly and voluntarily decide to mediate the dispute without any outside pressure, and set up the entire process themselves.

(a) The Mediation Process

Mediation is fairly simple, typically following a five-step process. Each of the steps helps to organize and structure the negotiation to

---

[1] Note that just because information is disclosed in mediation does not mean it cannot be used later. Any information that can be obtained through normal litigation channels, such as production and discovery, can be used. Only communications solely available through the mediation, such as settlement offers or statements made by any party during the process, are privileged.

maximize the opportunity for the parties to reach a settlement. The five steps are:

1. Pre-mediation;
2. Introduction and Opening Statements;
3. Identifying the Issues;
4. Exploring the Issues and Problem Solving; and,
5. Reaching and Documenting an Agreement.

(i)   Step 1: Pre-mediation

The pre-mediation step covers all actions taken before the parties actually meet, and ranges from quite simple to fairly involved. Some of the main considerations in the pre-mediation phase are:

1. **Choosing the mediator**: In some cases, the contract stipulates the mediator by name, in other cases a neutral organization such as the ADR Institute of Canada is named to assist the parties in appointing a mediator. In any event, the parties are always free to simply agree between themselves who will be appointed as mediator. Once a mediator is chosen, the mediator will typically organize and manage the other pre-mediation steps.
2. **Mediation briefs**: Prior to meeting, the parties prepare mediation "briefs", which are typically a summary of the issues as each party sees them, along with each parties' view of the dispute and what they want in order to resolve it.
3. **Exchange of important documents**: Parties must agree on what documents need to be exchanged prior to the mediation and typically they attach these documents to the mediation brief. If neither party has seen the other's expert reports, for example, valuable mediation time will be wasted while those reports are reviewed and assessed. The mediator will orchestrate the exchange of documents ahead of time to ensure that the mediation is meaningful and effective.
4. **Who is attending**: Exactly who is attending from each party can greatly influence the success of the mediation. If there are personality conflicts between certain people, other representatives should be considered. If there are significant engineering issues on the table, the engineer or architect should consider attending. Someone with adequate authority must also attend. All of these issues should be sorted out prior to the mediation.
5. **The Agreement to Mediate**: This is a written agreement that all parties sign prior to coming to mediation to ensure that everyone agrees with the principles and process of mediation. The agreement is a contract among the parties that commits them to abide by the terms of the process, and covers at a minimum the following issues:

(a) **Role of the Mediator**: The mediator is a neutral facilitator who will not give legal advice to any party.
(b) **Authority**: Parties confirm they will have authority to reach a resolution at the mediation.
(c) **Confidentiality**: The mediator commits to full confidentiality, and the parties commit to keeping everything from the mediation process confidential (except what can be obtained through normal channels for other processes, such as arbitration or litigation). Parties agree the process is legally without prejudice. The parties agree not to call the mediator as a witness under any circumstances.
(d) **Disclosure of Information**: Parties agree to full disclosure of relevant information to assist with settlement.
(e) **Voluntary Process**: Parties agree that the process is voluntary, and any party, or the mediator, can terminate the process if it is not working for them. Parties commit to discussing their reasons for terminating with the mediator before leaving.
(f) **Resolutions are Binding**: Parties agree that once an agreement is reached, it will be binding on all parties.

(ii) Step 2: Introduction and Opening Statements

The mediator convenes the first session, and briefly covers a number of issues at the beginning of the first joint meeting. The mediator conducts introductions of the parties, explains the mediation process and everyone's roles and expectations, and establishes basic ground rules for the meetings. These ground rules typically include things like one person speaking at a time, and respectful communications throughout the mediation. The mediator confirms that parties have the authority to settle, and describes further ground rules about how caucus, or private meetings with the mediator will be run. The mediator confirms everyone is comfortable with the process itself. Once everyone is comfortable with the process itself, the mediator ensures that everyone has signed the Agreement to Mediate[2] discussed above.

Once the process has been agreed to, the parties have individual opportunities to make an opening statement that summarizes their views of the dispute, the issues that are in dispute, why they see the issues the way they do, and what they want out of the situation. These statements are often summaries of what has been stated in the briefs, and can be given by either the lawyer or the parties themselves. Frequently, counsel makes the opening statements, but the client is encouraged to add his or her comments and

---

[2] A sample Agreement to mediate is included in Appendix 2.

perspective. This allows the client to be directly involved with the mediation, and allows each side to assess the credibility of the other party. Finally, it allows the mediator to define and clarify the issues that need to be resolved.

(iii) Step 3: Identifying the Issues

This is a critical step in the mediation process. Each party has presented its view of the dispute, and in most cases each party has omitted information that does not fit well with its own point of view. The mediator works with the parties jointly to "fill in the blanks", to help both parties get all of the information clearly on the table in relation to each issue, and to allow each party to ask questions of the other. By going deeper into each issue and generating a complete list of issues that need to be resolved, the mediator helps structure the negotiations effectively.

(iv) Step 4: Exploring the Issues and Problem Solving

The first step in the problem solving process is the mediator working with the parties to uncover and clarify their interests around each issue, their wants, needs and concerns. This is a critical step that helps the parties move away from a narrow focus on their "rights" or their "positions" to begin looking at the larger picture of what they really want and need in order to resolve the problem. As the mediator works with the parties to explore the issues and uncover the interests, parties are often alternating between joint or plenary meetings and caucuses or separate meetings with the mediator.

Essentially, this step is where the rubber really meets the road. Once the parties have clearly identified the issues that need resolution, and once they have revealed and discussed their interests, it is time to start looking at options for resolution. This process can happen in a few different ways.

First, the mediator might help the parties develop criteria or standards for assessing options and solutions. These criteria may include things like speed of resolution, respect of contractual rights, cost of reaching resolution, future working relationship, financial impact, *etc*. By building joint criteria for evaluating any options discussed, it helps keep the settlement based on objective standards rather than emotion.

Second, the mediator might encourage the parties to brainstorm options for settlement. A key part of this is encouraging the parties to think outside the box. Frequently, parties frame disputes as simply a money dispute, when in fact many solutions to problems can be far more creative.

When looking at a full range of the parties' interests, unexpected solutions frequently pop up.

Finally, the mediator helps the parties clarify and refine the solutions until a solution that works for everyone is agreed to. This can be done in a number of ways. It can be done jointly, with all parties in the room modifying and refining a "single text" solution. It can be done in the traditional offer/counter-offer format, where the mediator takes an offer from one party to another, who counters that offer with its own offer. The mediator continues this until the offers converge. A third way is to have each party rank the most promising options along with his or her reasons, slowly striking low-ranked options off the list until only one remains. The choice of which process to use depends on many factors, such as the relationship between the parties, the nature of the dispute itself, and the style of the mediator.

(v) Step 5: Reaching and Documenting an Agreement

Once an agreement has been reached in principle, a final step must be taken to document the terms of the agreement that covers all aspects of the settlement. Construction disputes can be very complex, and the solutions to those disputes can be equally complex. Many times, once parties reach an agreement in principle they feel the problem is solved and want leave without properly documenting the resolution. This is a recipe for creating an even larger dispute in the future.

The mediator will usually insist the parties sit down and draft a clear document, typically called "Minutes of Settlement", that will contain all relevant terms of the agreement clearly spelled out. In some cases the mediator will help the parties draft this agreement, in other cases the mediator will require counsel for the parties to do the drafting. Either way, the parties leave with a binding resolution to the dispute.

(b) Substantive Knowledge of the Mediator

When choosing a mediator, there is an important question that parties must answer about how much expertise in construction the mediator should have. Some people believe that mediation skills are generic enough that regardless of any substantive knowledge, the mediator can effectively help the parties. The opposing view, however, is that due to the complexity of construction disputes, substantive knowledge of some degree is a must to be effective as a mediator.

In reality, parties should look for an effective balance of both strong mediation skills combined with a knowledge and understanding of the construction field. First and foremost, parties should know that the mediator is indeed skilled as a mediator, as those are the essential skills that will help the parties negotiate effectively. With the growth of mediation, however, there are many individuals who bring a strong background in construction along with good mediation skills to the table. Note that it is not necessary that the mediator is an absolute expert in all aspects of construction, but the mediator does need a strong general knowledge of construction issues and the construction process. This view is well summarized by Dwight Golann, a construction lawyer turned mediator, who said: "Subject matter knowledge is always desirable in a neutral, provided that he or she has strong process skills".[3]

This view is echoed by the Canadian Construction Documents Committee rules for mediation, which states: "The Project Mediator must be impartial and independent of the parties and an experienced and skilled commercial mediator...and have knowledge of the relevant construction industry issues".

Strong knowledge of the construction industry combined with good mediation experience and skills is an ideal combination.

(c) Pros and Cons of Mediation

The greatest strength of mediation as a dispute resolution process is the way it shifts the parties' focus away from winning or beating the other side, and refocuses the parties on clarifying and understanding what they really want to resolve the dispute. Conflicts frequently follow a pattern of escalation, where initially Party A wants the problem fixed, plain and simple. When nothing happens, Party A still wants it fixed, and now wants Party B to also acknowledge that it was not Party A's fault. When this does not happen, Party A moves to wanting the problem fixed, plus all damages he or she is legally entitled to (and maybe more), plus a public apology that absolves Party A completely. When this does not happen, Party A wants all of the above, plus he or she wants Party B punished to boot. When parties show up for dispute resolution, they are frequently in the last stage of escalation.

Adversarial processes such as litigation and arbitration tend to reinforce the desire for punishment and revenge, because they emphasize a third-party determination of right and wrong, of blame and fault. Media-

---

[3] D. Golann, "Mediating Legal Disputes: Effective Strategies for Lawyers and Mediators" *Aspen Law and Business* (2001), at 36.

tion on the other hand, focuses parties back toward the early steps of the escalation, refocuses parties on what will fix the problem and what will minimize the costs and damage to the relationships going forward. This de-escalation and focus away from blame and revenge frequently helps parties arrive at settlements they can live with. In other words, construction mediation is a way to dissolve the "conflict poison" by reaching an agreement that is fair to the parties in an atmosphere of co-operation and mutual respect.

In addition to the above, some of the other advantages to mediation include:

- Low cost: Compare one or two days of mediation to 2 or 3 years of litigation, or even six months of arbitration, and you can see the significant cost advantages.
- Confidential: Like arbitration, mediation is confidential, and helps parties quietly resolve the conflict and move on.
- Creative outcomes: Mediated solutions can help parties jointly think outside the box, and find solutions that could never be found in litigation or arbitration. This is primarily because mediation is collaborative, where litigation and arbitration are adversarial. It is extremely difficult to be creative when you are trashing and bashing the other party, and vice versa.
- Low risk: By mediating, parties do not give up any of their legal rights, and do not prejudice their claim in any way. Parties can continue on the arbitration or litigation path not only without any risk, but also with a better understanding of the dispute.

Mediation of construction disputes is particularly effective because the parties' interests are rarely limited to just legal or monetary issues. Unlike litigation and arbitration, mediation can address and resolve a much wider range of interests, including non-monetary solutions and future working relationships. As Thomas Crowley comments in his recent book: "Construction projects especially lend themselves to mediation, because the parties usually have mutual needs for maintaining their business relationship and obtaining expertise from a neutral party to help settle the dispute."[4]

Mediation, however, is not a panacea and has some disadvantages, too. Some of the disadvantages include:

---

[4] T.E. Crowley, *Settle It Out of Court: How to Resolve Business and Personal Disputes using Mediation, Arbitration and Negotiation*, John Wiley & Sons Inc., 1994, at 55.

- Closure: Mediation does not necessarily bring closure. With roughly an 80 per cent – 85 per cent settlement rate the odds of settlement are high, but not guaranteed.
- Additional costs: While mediation is relatively inexpensive, it is not free, and it does add time and cost to the process, something that parties should look at carefully.
- Strategic risk: In some cases, even though the process is confidential, parties can get a better understanding of the other side's case through mediation. This can help the party to simply try and strengthen their own case before trial, rather than try to settle.
- Delay: If the other party has no intention of settling and is using mediation as a delay tactic, they can stretch the dispute out a few additional months by going through the motions of mediation.
- Precedent: If the goal of one party is to set a legal precedent, mediation is not the place to do that.

Overall, mediation represents a low cost, low risk opportunity to resolve a dispute quickly and effectively, and should at least be considered for every dispute. It is because of this low risk that most experts recommend "presumptive mediation", meaning that we should ask the question, "Why *shouldn't* we mediate?" rather than asking, "Why *should* we mediate?"

II. NEUTRAL EVALUATION

Neutral evaluation is a generic term for any process where the parties receive a non-binding evaluation of the dispute from a neutral third party. Neutral evaluation is a blend of interest-based and rights-based approaches, interest-based in that the results are non-binding, and rights-based since the evaluation of the parties' positions focuses significantly on the parties' rights, on what each is entitled to should the claim proceed to court. What often distinguishes the different approaches to neutral evaluation are questions like who is doing the evaluation, how much time and effort are the parties putting into presenting their side of the dispute to the neutral, and what happens after the evaluation to help the parties reach a resolution. There are a variety of different approaches to neutral evaluation ranging from simple and fast to involved and detailed. Three options are described below that cover the full range, along with some of their advantages and disadvantages.

(a) Neutral Expert Opinion

This is a relatively informal process where the parties mutually agree to engage an industry expert to review the problem and render an opinion. This review can be simply an opinion about what the defect or problem itself is, *i.e.*, *what* caused it, or it can be a finding and apportionment of fault or blame, *i.e.*, *who* caused it. The parties need to decide up front exactly what kind of opinion they want from the expert.

The advantages to a neutral expert opinion are that it is the least formal of the neutral evaluation processes, does not require formal hearings or even legal representation, and can be done quickly, effectively and cheaply. It is most effective when there are substantive disagreements about what actually happened, or what contributed most to the disputed issue. A great deal of the value of this type of evaluation rests on the credibility and expertise of the neutral. The disadvantages to this type of evaluation include the fact that it may be hard for parties to agree on an expert or to accept the expert's finding of blame, given the informal and less structured approach to hearing the issues.

(b) Non-Binding Arbitration

This is a much more formal, structured process and has many of the same procedural issues that binding arbitration has, detailed in the next section. In non-binding arbitration a formal hearing is conducted and a full finding of fault with a recommended award is given to the parties for them to consider, but the award itself is not binding. There are variations, such as paper-based or documentary arbitration where the arbitrator works solely from documents without a formal hearing or testimony of witnesses.[5]

The advantage to non-binding arbitration is that the evaluation or outcome is the result of a full hearing, and therefore may carry more weight with the parties than an expert opinion. The disadvantages include the fact that it is costly, takes longer, probably has counsel involved, but does not necessarily end the dispute. After the non-binding award is released, the intent is for the parties to continue negotiating based on the arbitrator's opinion, but at times the nature of the award itself can prevent further negotiations. For example, if the decision is 100 per cent in favour of one party, the other party may withdraw to strengthen its case rather

---

[5] While being less expensive, the paper-only approach not only limits the time and costs of the process, but also the perceived quality of the decision.

than negotiate from a position of weakness, causing more delays than resolution.

(c) Mini-trial

The mini-trial is probably the most elaborate form of neutral evaluation. In a mini-trial, parties conduct a full mock trial within a fixed period of time, typically in the one- to three-day range. These mini-trials are conducted in front of a "judge", a neutral who hears the case and issues a verdict at the end of the trial.[6] Also attending the trial are the CEOs of both disputing parties, who observe the trial, review the verdict, and then formally begin negotiating a resolution to the issue. In some cases the neutral can shift roles and mediate the dispute with the CEOs.

In this process the parties get the full flavour of a trial, but condensed into a short period of time. They get a decision with reasons to consider; the negotiations that follow are then conducted not by the people who were involved in the problem from the beginning, but by the CEOs (or other senior managers) who come to the problem with a fresh pair of eyes. The disadvantages to mini-trials are the costs and time involved, and because of this mini-trials are only used in very significant cases. Similar to non-binding arbitration, mini-trials do not guarantee a final solution, in spite of the time and money spent.

III. ARBITRATION

Arbitration is essentially a process for having a dispute privately adjudicated. It is a non-judicial legal technique for resolving disputes by referring them to a neutral party for a binding decision, or "award". An arbitrator may consist of a single person or an arbitration board, usually consisting of three members. The parties themselves hire the neutral of their choice to hear the opposing sides of the dispute, and at the end of the hearing the arbitrator(s) render a final and binding judgment. There are a number of important features of the arbitration process, including:

- Arbitration follows similar rules of evidence as litigation, and as a result is a highly structured and formal process that resembles the litigation process in many ways.
- The parties can jointly choose the arbitrator or arbitrators, and can therefore choose someone with direct knowledge and expertise in the

---

[6] In more elaborate forms of mini-trials, the parties actually empanel a jury, which deliberates and renders a verdict.

construction field. The trade-off for the ability to choose the arbitrator is that parties typically give up any right to appeal the arbitrator's award.
- Arbitration can be done either by a single arbitrator, or by a panel of three arbitrators. Where a panel is used, each side typically appoints one arbitrator, and the two arbitrators choose the chair of the panel to bring the total to three.
- Parties can have some input into the structure of the hearings and the way the arbitration is run, but the final decisions on process rest with the arbitrator.
- Since arbitration is adversarial in nature and similar in kind to litigation, counsel usually represents parties.
- The arbitration award is typically final and binding, with no recourse or appeal process of any kind.

(a) Pitfalls of Arbitration

Since arbitration shares many of the same features as litigation including rules of evidence, adversarial presentations to an adjudicator, proof of damages, *etc.*, it also shares many of the pitfalls. One of the top construction lawyers in the United States, John W. Hinchey, past chair of the American Bar Association's Forum on the Construction Industry was quoted as saying, "The construction industry is extremely frustrated with the legal profession, the judicial system, and of course litigation. Nor are they particularly infatuated with arbitration, which they think the lawyers have taken over and botched, just like the courts. To the extent possible, the construction industry wants to march into the 21st century without lawyers in their lives."[7]

The risk of arbitration, simply put, is that it will replicate the cost, time delay and adversarial quality of litigation without giving the parties an opportunity for appeal. The further that arbitration becomes "legalized", the less reason there is for parties to choose it. Before choosing arbitration, parties should consider some of the major concerns that have been expressed about arbitration for construction matters.

(i) Costs and Time

One of the main dissatisfactions with litigation of construction disputes (and indeed of all disputes) is the length of time litigation takes and

---

[7] J.W. Hinchey, "A Wake-Up Call to the Construction Bar" (1997) 2 Construction Forum News, No. 2, at 1-2.

the costs involved. In theory, arbitration delivers better results on both of these fronts, and indeed in many cases, this is true. It is not, however, a guarantee, and there are many situations where arbitration takes longer and costs more than litigation might have. As stated by Stanley Naftolin, J.D., Q.C., current president of the Canadian College of Construction Lawyers, "Compared to litigation where the case proceeds through the civil court system, arbitration can be and usually is much quicker, but the cost of the arbitration is not any less expensive and may, in some circumstances, be more expensive."

In a recent example, a retired judge was sitting as the chair of a panel of three arbitrators. One of the parties was taking a great deal of time presenting cumulative and repetitive evidence as part of its case, which the other party objected to. When the arbitrators convened to discuss requiring the long-winded party to focus and shorten its presentation, the chair remarked that if this had happened in his courtroom he would absolutely not put up with it, but since this was arbitration, it was "their dollar, so let them take as much time as they want."

(ii) Fairness and Predictability of Outcome

One of the arguments for arbitration over litigation is that because parties can jointly choose the arbitrator or arbitrators, they can choose arbitrators with significant subject matter expertise. By doing this, the parties will receive decisions that are fairer, and reflect an outcome based on the actual reality of the construction industry. While it is certainly true that in many cases the arbitrator's knowledge of construction processes will help, it is naïve to think that this knowledge will necessarily mean that the outcome will be any fairer or more predictable than a court decision made by a judge with only passing knowledge of the construction industry.

In 2002, the Construction Management Association of America's CM eJournal[8] reported a survey of 2,000 construction industry practitioners. The survey requested respondents to assume the role of a well-respected arbitrator appointed by the three parties to a dispute. They were given a detailed summary of the facts, testimony and information presented at the hearing, and asked to fill out a questionnaire that included detailing what they would have awarded to which party in the dispute. The questionnaire did not limit the scope or the amount they could award, and it included a request for the reasoning and explanation for the award.

---

[8] George (Chip) Ossman III, "Construction Arbitration: A Survey to Arbitrator's Award Consistency" (2002) Construction Management Association of America CM eJournal.

The fact scenario was a delay claim between an owner, a general contractor, and one of the general contractor's subcontractors. Both the general contractor and the owner agreed that the subcontractor was *not* at fault for the delay. The claim summary was:

- The subcontractor was claiming $60,000 against the owner and/or general for a 30-day delay;
- The general contractor was claiming $90,000 against the owner for the delay, and denying responsibility for the subcontractor's losses;
- The owner was claiming $150,000 against the general, and denying responsibility for the subcontractor's losses.

430 of the 2,000 responded to the survey. The results of the survey were as follows:

- 94 per cent of awards determined the subcontractor should receive the full $60,000 claimed.
- 34 per cent determined the owner and general should each pay $30,000 to the subcontractor, with no further costs paid (a "split the baby" award).
- 35 per cent determined that the owner was responsible for the full $60,000 to the subcontractor. In addition, 22 per cent of these respondents ordered the owner to pay additional costs to the general contractor.
- 18 per cent determined that the general contractor was responsible for the full $60,000 to the subcontractor. Six per cent of these respondents ordered the owner to pay additional costs to the owner.
- 13 per cent found a variety of intermediate awards that ranged all over the map.
- Four per cent determined that the owner should pay the maximum, $60,000 to the sub, and $90,000 to the general.
- Less than one per cent decided the general contractor should pay the maximum, $60,000 to the sub, and $150,000 to the owner.

While these results are certainly open to interpretation, a few outcomes jump out at the reader. First, there is very strong agreement that the subcontractor should receive all its money, in large part no doubt to the general and the owner agreeing that the sub was not at fault. But in spite of that agreement, 6 per cent of arbitrators still did not give full relief to the subcontractor.

The results as between the owner and the general contractor range completely across the board, with a small bias in favour of the general. Thirty-four per cent of arbitrators simply "cut the baby in half", a frequent criticism of arbitration in general. Of the remaining responses, twice as many gave strong awards in favour of the general than the owner. That said, it should be clear that this was a no-win situation for both the general and the owner. The owner had only an 18 per cent chance of "winning", while the general had only a 34 per cent chance of "winning" — both would have been better off flipping a coin!

In the final analysis, arbitration outcomes, even with experienced and knowledgeable arbitrators, can be just as unpredictable as the outcomes in litigation.

(iii) Lack of Right to Appeal

Parties entering arbitration give up their right to appeal on almost any basis, including errors of fact, law, or mixed fact and law. Courts are loath to overturn arbitral awards, for the simple reason that in litigation you cannot choose your judge, nor can you necessarily get a judge with an understanding or expertise in the area of law your case is from. Therefore, a decision that is made by a judge or jury must have an avenue of appeal in order to achieve due process.

In arbitration, however, parties are free to choose their arbitrator. They can jointly select an individual that they have confidence in, and that has the expertise and knowledge to fairly decide the case. For this reason, most statutes governing arbitration enshrine the final and binding nature of the awards, and the courts generally refuse to second-guess the arbitrator who has been selected and agreed to by the parties. The only exception to this is where a decision is clearly tainted by a newly discovered conflict of interest (where the arbitrator is related to one of the parties, for example) or where "natural justice" has been breached. Those are very rare situations, so parties going into arbitration need to be aware that the decision is for all intents and purposes final and binding.

While there are significant pitfalls to the use of arbitration in the construction environment, it has been and can continue to be a useful tool in resolving disputes. In the next section, we will look at best practices in arbitration that minimize the pitfalls discussed.

IV. BEST PRACTICES IN ARBITRATION

Construction projects can be some of the most complicated, detailed, difficult and expensive processes humans have invented. In spite of this, many projects run smoothly and effectively, producing outstanding results at the end. The reason for these successes, when they occur, is because they use state-of-the-art planning, scheduling and project management techniques. By applying many of the same process management techniques used by the construction industry itself, the arbitration process can also deliver effective results for the parties. When considering arbitration, parties should explore and recommend the use of some or all of the following best practices:

(a) Strong Initial Focus on the Procedural Side of the Arbitration

This step is critical. The arbitrator should take firm hold of the process, and in consultation with the parties, discuss and reach agreement on the following:
- A realistic assessment of the time needed for discovery and preparation, and for the hearing itself;
- The pre-hearing dates should be set aside in advance in anticipation of unexpected issues arising;
- Early identification of the issues, and a focus on narrowing the issues well in advance;
- A limit on the amount of hearing time that will be used, and have the parties fit the presentation of their case into the amount of time allotted;
- Procedural goals for the arbitration, including timing and deadlines. Ask the arbitrator to stick to those goals;
- Facilitation of early and open exchanges of information and documents needed for the hearing;
- Schedule and exchange expert reports early in the process;
- Identification of witnesses and their experience and qualifications well in advance of the hearings.

(b) Strong Focus on the Handling of Evidence to Minimize Time and Delays

There are a number of steps that can be required by the arbitrator to maximize the effectiveness of the hearing, including:
- Make use of common books of facta, such as contract documents, meeting minutes, change orders, exchange of letters or email, *etc*.;
- Create well organized and numbered exhibits, listed chronologically;
- Allow self-authentication of document copies, with a right to object on a case-by-case basis only;
- Convene on site if it will help the arbitrator understand the context;
- Require all documentary evidence to be on computer in searchable format whenever possible to minimize time spent searching for documents;

- Encourage use of visual aids, such as graphs, charts, photographs, videos, *etc.*;
- When opposing experts will testify, require them to meet ahead of time to identify areas where they agree and where they disagree. Have them testify jointly to answer questions at the same time.

(c) Tight Control of the Hearing by the Arbitrator

Choose an arbitrator who will take a firm hand in minimizing the parties' use of technical/legal objections, repetitious evidence, abuse of process or delaying tactics.

By considering and following the processes above, significant benefit can be achieved for the parties should they choose the route of arbitration. In one case, for example, an arbitration was convened on March 27th in a dispute involving the government, four days before the government's fiscal year-end. With the consent of the parties, the hearing was conducted and the award released by March 31st so the result could be booked in the current fiscal year. Arbitration, when properly set up and effectively run, can be an excellent alternative to litigation.

V. HYBRID PROCESSES

There are a number of hybrid dispute resolution processes parties can turn to that combine different features and approaches. Many, however, are simply variations on the processes already mentioned, so we will look at the two most common hybrid processes, med-arb and arb-med.

(a) Med-Arb

Med-arb refers to the combining of the mediation process and the arbitration process into one. In a typical med-arb the neutral third party begins the process as a mediator, facilitating the negotiations between the parties with the goal of resolving all issues through mediation. Should settlement not occur, however, the mediator simply changes hats and becomes an arbitrator. At that point, the arbitrator is given great freedom to either immediately issue an award on all outstanding issues, or to hold further hearings into the outstanding issues before issuing an award. In either case, however, the award, just like in regular arbitration, is final and binding on the parties.

The reason the med-arb process is chosen is twofold. First, parties are looking for finality and resolution, and mediation alone does not guarantee this. By adding the role of arbitrator to that of mediator, finality is guaranteed one way or another. The second reason is one of costs. If the mediator spends a few days or weeks with the parties helping them negotiate but parties fail to reach a resolution, starting over with a new person as arbitrator means that all the time spent by the mediator learning about the issues is lost. In theory, the mediator brings along all that knowledge when he or she becomes the arbitrator, and this saves time and money for the parties.

There are a few significant issues with med-arb, the main one being the parties' anticipation of the "transformation" from mediation to arbitration. In other words, if parties know that the mediator might be deciding the issues later, they will treat the entire mediation process simply as arbitration, and not seriously negotiate to resolve it. In addition, they may not be completely open and honest with the mediator, because they are afraid that if they admit some fault to the mediator, the arbitrator (formerly mediator) will hold that against them later when issuing an award. It can promote parties taking extreme positions, which defeats the purpose of the mediation process.

(b) Arb-Med

As a way of minimizing the problems with med-arb, a second hybrid process was developed called arb-med, which simply reverses the two roles that the same neutral plays. In arb-med, the neutral first holds an arbitration hearing, listens to the case, and issues an award in a sealed envelope without sharing it with the parties. The arbitrator, who is now very familiar with the issues, then mediates the dispute with the parties. Should the parties reach a full and final negotiated settlement, the award is simply destroyed without being revealed. No one will know what the outcome would have been. Should the parties not settle the dispute, the envelope is opened and the award is final and binding. None of the revelations of the parties made during mediation will have affected the arbitrator's judgment.

Arb-med is a clever way of helping the parties reach resolution, as the parties have a chance for their case to be heard, as well as a chance to see and assess how well the hearing went for them. Then, they have a chance of resolving it collaboratively. The main disadvantage of arb-med is again during the mediation process. Once the decision has been made and the mediator knows the outcome if they do not settle, parties may spend a lot of time "reading into" everything the mediator says, trying to divine who the winner and loser might be. If either party is doing this rather than negotiating, then the process is likely to fail.

VI. LITIGATION

The many weaknesses (and strengths) of litigation have been written about endlessly elsewhere, so we will confine the discussion here to a few specific areas, such as costs and complexity.

(a) Costs

It would be hard to find many litigants at the end of the litigation process that are surprised with how little it cost. In fact, it is a constant source of surprise at how expensive litigating anything, let alone complex disputes in the construction industry, can be.

Ten years ago, the Attorney General's office in Ontario formed an advisory group to look at ADR for construction disputes.[9] In a report issued in June of 1994, the advisory group produced an estimate of the costs of litigating a construction matter. In summary, they found the following:

Legal Costs of a $100,000 Construction Claim

Activity:	Days:	Client Time:
Interview and Draft Pleadings	4	2
Affidavit of Documents	3	5
Witness Interviews	4	2
Discovery	11	7
Interlocutory Motions	10	1
Preparation of Experts	4	0
Pretrial	3	0
Preparation for Trial	10	5
Trial	12	12
Total Lawyer Days	**61 Days**	**34 Days**
Total Lawyer Costs:	$140,000.00	

---

[9] Advisory Committee on the Alternative Resolution of Construction Disputes, *Too Many Disputes! Too Much Litigation! Dispute Resolution Opportunities for the Construction Industry* (Province of Ontario, Ministry of the Attorney General, June, 1994).

Disbursements:	$6,000.00	
Expert Fees:	$15,000.00	
Total Legal Costs:	**$161,000.00**	

Note that parties will each be spending over $160,000 in legal costs, along with over 30 days of lost opportunity costs of their own time in order to litigate who gets $100,000! Most businesspeople would not invest in a business with these prospects.

If anyone thinks that the above costs are exaggerated, they need only to look at *Foundation Co. of Canada v. United Grain Growers Ltd.*[10] As summarized by construction attorney Geza Banfai, in this case:

> "...the combatants engaged in 99 days of examination for discovery plus another 132 days of trial, spending over $5 million in legal fees among them, to reallocate approximately $1.6 million in claims amongst themselves."[11]

It should be noted that at the time Banfai wrote the above, the case was under appeal, and the parties were still spending money.

(b) Complexity

As noted in the arbitration section, the significant risk in litigating construction disputes is the fact that few judges or juries have a clear understanding of the construction field or the construction process. Consequently, parties have to assess the quality and predictability of the decisions that they are likely to receive in a court setting. In addition, the complexity of the dispute and the court process itself often results in lengthy trials, as each side feels it must educate the judge in all aspects of its claim to get the best chance at winning. All of that adds to the cost and the length of time to get a result.

The best practice around litigation in the construction industry is, quite simply, to avoid it, but to avoid it effectively through some of the other best practices discussed in this book.

---

[10] (1995), 25 C.L.R. (2d) 1 (B.C.S.C.), additional reasons at (1996), 25 C.L.R. (2d) 1 at 204, 8 C.P.C. (4th) 354, [1996] B.C.J. No. 2090 (S.C.), revd in part (1997), 91 B.C.A.C. 254, 148 W.A.C. 254, 34 B.C.L.R. (3d) 92, [1996] B.C.J. No. 969 (C.A.).

[11] Geza Banfai, "Why Construction Lawyers Must Change", 36 C.L.R. (2d) (1998). This is a revised text of an address delivered at a meeting of the Allied Professions Committee of the Toronto Construction Association on June 10, 1997.

VII. SUMMARY

This chapter has reviewed a full range of dispute resolution processes that the parties can use to address disputes that remain at the end of the project. In dealing with disputes at the end of the project, it is important for parties to take proactive action to address them, and resist the temptation to put them "out of sight, out of mind" in the litigation stream. While this approach takes less energy at the time, it frequently results in significant costs down the road.

Disputes can be either a source of cost, expense and aggravation, or they can be a chance to work toward resolution and rebuild relationships in the process. By choosing the right dispute resolution process, parties can improve their outcomes substantially.

In the next chapter, we will look specifically at mediation, at the most effective way to prepare for mediation in the construction industry.

VIII. CASE STUDIES

(a) Case #1: If You Can't Stand the Heat, Get Out of the Kitchen

This was a construction dispute between the owner of a new high-end custom built home, the builder, and the manufacturer of the kitchen cabinets.

The home was of very high-end quality and the owners were willing to pay for what they wanted. They had ordered and purchased a kitchen from the builder's kitchen supplier. The kitchen cost was over $70,000.

All of the proper documentation was filled out including the style, material, finish, and colour of stain to be used. The kitchen was delivered and installed on time and everything seemed to be in order. The owner stated that when they took possession of their home they completed the pre-delivery inspection with the builder. They documented some deficiencies of a minor nature but were very pleased overall with the high-end quality of their new home.

A month after moving in they notified the builder that the kitchen cupboards' finish had started to bubble. The builder came to have a look at the problem and agreed that something was wrong. The builder notified the kitchen manufacturer who also said it would come to look at this problem. When the kitchen manufacturer came to look at the problem, it also agreed that something was wrong. The manufacturer's representative took a cupboard door with him back to the plant to have it analyzed.

A month went by without a response from anyone so the owner followed up with calls to the kitchen manufacturer. After two weeks of leaving numerous messages, the owners went to the kitchen manufacturer's retail outlet where they had purchased their kitchen. The company no longer employed the salesperson they worked with. They asked to see the representative who had come to their home and taken the cupboard door for analysis. They were told the representative was out and they left a message for him. Again, they had no response.

Another month went by and the owners, starting to get frustrated, began to leave angry and threatening voice messages. Finally they received a return call from the president of the kitchen manufacturer. There were told that the problem was with the stain, and they were in discussions with the stain manufacturer to see if they would cover the cost of redoing the kitchen. The owners asked when this problem would be rectified, and were told that it would be dealt with within a week.

Trust had now diminished substantially. To make sure it would be handled, the owners contacted the builder who seemed surprised that this problem had not been resolved yet. He told the owners that he would deal with this issue himself and get back to them within an hour. Another week went by with no response.

The owners, now completely exasperated, got in touch with their lawyer. The lawyer explained to them the different ways they could proceed in order to have the problem resolved. The lawyer also wrote letters to the parties informing them of possible litigation. That seemed to get some action, as the parties all responded to the lawyer's letter, apologizing for the delays in getting back to the owners. They proposed a meeting at the owners' home to discuss and implement an action plan. The owners agreed, but followed their lawyer's suggestion of bringing in a facilitator to help analyze and resolve the situation. The facilitator was a builder the lawyer knew who mediated in the field. All the parties agreed.

The meeting was held at the owners' home two days later. As soon as the meeting began, the owners became angry and rude, personally attacking both the builder and the manufacturer. The builder and manufacturer got up to leave, not willing to be treated that way. The facilitator stepped in, and took the owners to another room, asking the builder and manufacturer to wait for him. The facilitator asked the owners what was upsetting them so much, and found out that what made them so angry was being ignored, and treated like their problem did not matter. This, after spending $70,000 on the kitchen! In addition, none of the other deficiencies had been repaired either, nearly three months after completion. They had paid a lot of money, and wanted the quality they had been promised.

The facilitator brought everyone back together, and had the owners, now somewhat calmed down, explain why they were so upset. The response surprised the owners. The manufacturer apologized, saying they had been so busy that a number of things had slipped through the cracks and they did not realize how much time had gone by. The builder said that he was also at fault for not following up. He explained that because of the increased demand for his services he had fallen behind. Both parties apologized to the owners and assured them that things would be dealt with immediately.

The next step was to agree and implement an action plan to resolve the problem. The kitchen manufacturer explained that it had been waiting for the stain company to contribute to the cost. Since that had not happened, it had gone ahead with a new kitchen using a different stain, and it would be ready for installation in less than a week. The builder also committed to having all deficiencies repaired within the week, and made it clear to the manufacturer that if it were not installed on time, it would jeopardize their future business relationship.

The owners relaxed, happy to finally receive an apology and some focused attention on their problem. As the meeting wrapped up, they asked the kitchen manufacturer what it would cost to buy the old cabinets for their cottage, in "as-is" condition. The manufacturer promised to give them a great deal. The meeting ended on a positive note for everyone.

(i) Lessons from the Case

1. **Dealing with Emotion**: Facilitation and mediation are exceptionally good at helping parties manage strong emotions. This situation could easily have escalated to litigation because of the strong emotions. Instead, the facilitator helped guide the parties through the strong emotions, and kept the discussions on track.
2. **Intervene Early**: There is a tendency to ignore problems that are not yet "burning". This is a mistake. By intervening earlier, most of these issues and problems could have been avoided.

(b) Case #2: Pulling the Plug

The following case points out how extremely poor relationships can prevent a final resolution, and how mediation can assist in achieving partial solutions even in very difficult situations.

This was a four-party construction liability case that included a power utility as the plaintiff, two defendants (a contractor hired by a municipality to replace a water main and a utility-locating service), and one third party (the municipality itself). The defendants and the third party had all cross-claimed against each other.

The situation was fairly straightforward. The utility-locating company had been called in by the contractor, had marked a number of underground utilities, and had expressly stated that the contractor was required to hand-dig within 3 feet of any markings. They had marked two utilities crossing the road. The contractor had hand-dug two holes on each side of the road, and located two utilities at a depth that was not in its way, had dropped a "torpedo" into the hole to tunnel under the road, and had severed a large power cable that ran very close to the two marked utilities, but at a shallower depth. Power was knocked out for blocks around, emergency generators set up, and major repairs had to be completed by the utility. The initial claim in the lawsuit was just over $500,000. The contractor claimed that the power line was not marked properly on the locate done by the locating company, and the locating company countered that the contractor was supposed to hand-dig regardless, and if they had, they would have found the cable. The contractor also claimed that the city should have known about the cable and should have included that information when it hired the contractor, thus making the city liable, too.

It was clear that communication between the contractor and the locate company throughout the project had been poor, and there had been other incidents where the locate company did not show up on time, or did not provide the level of information the contractor wanted. Both were frustrated even before this particular incident. The municipality was angry that it was included at all, especially in the cross-claims from the contractor, since the two had had a long working relationship on many projects previously.

Because of the strained relationships, parties were very reluctant to share information about the events leading up to the accident, and had refused to bring key individuals forward prior to the mediation.

A mediation was scheduled, and it became apparent that counsel for the locate company and counsel for the contractor were quite hostile with each other, which further blocked communication. The first day of mediation yielded little movement, but did focus the discussion around two key issues:

1. Parties developed a list of information that everyone needed to have before they could move forward, and
2. It was clear that the plaintiff was indeed blameless, and had documented the claim for $500,000. Even further, there were lost

revenue calculations that had not been included in the plaintiff's claim as of yet, and those were estimated at an additional $300,000 if the matter went forward.

A second day of mediation was arranged 4 weeks later, and parties agreed to exchange the information on the list prior to the session. At the second session, the hostility between counsel for the locate firm and the contractor had grown and they could not stay in the same room together. They remained in caucus for the balance of the mediation. In spite of this, a consensus emerged that the plaintiff should be paid and settled out, and a number was found ($450,000) that everyone agreed to, including the plaintiff. In addition, new information made it clear to the contractor that the city had very little, if any, exposure, and another consensus emerged that the city should be released. A major barrier remained, however, and that was that the two parties refused to pay the plaintiff any money as of today, as each was so convinced that the other was at fault.

With the help of the mediator and a few hours of discussion, the plaintiff agreed to settle for their amount, but it would defer being paid anything until a trial was held with the two remaining defendants litigated solely on the issue of liability, with the following terms:

- The plaintiff would defer payment interest-free for one year, after which interest at 8 per cent would accrue against the losing party;
- Neither defendant would assert liability against either the plaintiff or the city, but only against each other;
- Whoever was found liable would pay what it was ordered to pay at the time of judgment, regardless of whether either party appealed.

In the end, two of the four parties left the mediation settled. The remaining two had their liability capped (they did not face the prospect of losing an $800,000 potential claim plus interest and costs, which everyone estimated would top $1,000,000 easily if the plaintiff won). In addition, the trial was streamlined and simplified for the two remaining parties.

(i)   Lessons from the Case

1. **Poor Relationships**: Even though in this case the poor relationships between the contractor and the locate company caused the mediation to fail on the issue of liability, the mediation process still allowed the parties to focus the issues enough to reduce the number of parties, and cap the plaintiff's claim pending a final resolution by the courts.
2. **Improved Communication**: The relationships between the contractor and the locating company were so poor that without mediation, no negotiation could have taken place. Through mediation and the

involvement of a neutral third party, significant progress was made even though the parties could not stay in the same room together.
3. **Maintained Relationships**: Through the mediation process, the relationship between the city and the contractor was preserved when the contractor agreed with the locating company to let the city out.

6

THE SEVEN HABITS FOR SUCCESSFULLY MEDIATING CONSTRUCTION DISPUTES

We have covered a broad range of best practices in managing and resolving construction disputes including steps to take prior to starting a project, during the project, and after the project has been completed. In this chapter, we will look at the seven most important steps, the seven most important "habits" to practise to successfully mediate construction disputes. These are the key steps that most influence the successful resolution of a dispute, starting with preparation for the mediation right through to signing an agreement.

I. HABIT #1: MEDIATE EARLY

When a mediation takes place is a critical question of timing. The two broad choices are mediate as early as possible, early being virtually as soon as the dispute is recognized; and late, late being after the lawsuit is commenced, discoveries and expert reports are completed, and the trial is pending.

(a) Mediating Late

Once a dispute has entered the litigation system there is a strong tendency, if mediation is considered at all, to mediate late in the process, very close to trial, for the simple reason that this is the best time for lawyers to advise their clients on the legal issues and risks. Why? Because damages have been crystallized and maximum information has been obtained and assessed, allowing counsel to be most comfortable with giving their clients clear legal advice. Mediating that late in construction disputes however, may be the worst time for all parties involved. The longer the dispute goes on, the larger the damage claims tend to become. While this may seem obvious, what often is not considered is the magnitude of the potential consequential damages. In a dispute between one subtrade and the general contractor, if that subtrade liens the project and brings it to a halt, damages are mounting for every single trade affected. This in turn brings damages from the owner into play. The net result is that what may have started as a small or medium size claim turns into a major damages file simply because of the delay in addressing it. In addition, if the project is ongoing at the time of the dispute, mediating late may eliminate the chances of non-monetary resolutions such as early completion dates, upgrades, *etc.* By waiting, the parties' relationships also deteriorate, often causing their positions to harden. While it still may be better to mediate late than not mediate at all, there is real leverage in mediating as early as is reasonably possible.

(b) Mediating Early

If the project is ongoing at the time of the dispute, it should be a major priority to keep things moving ahead. By convening the parties while the dispute is still fresh, positions have not had as much time to harden and memories of events are clear. There is far less conflict poison to address early in the conflict, and this helps the parties look at resolving the issues quickly and moving on, rather than digging in for the long haul. Since the project is proceeding, solutions can be considered that fit in with the ongoing work, allowing for innovative solutions that make sense for all parties. In addition, by mediating one issue early other issues can be addressed at the same time, preventing related disputes down the road. While issues will not settle if addressed too early, if key data is not known or the full extent of the problem cannot yet be assessed, parties should take the risk of mediating too early rather than too late.

II. HABIT #2: PREPARE, PREPARE, PREPARE

Almost everything at mediation hinges on the parties' level and quality of preparation. Unlike many other legal disputes, construction disputes are often complex and multi-party, which means that the clarity and focus of the parties coming into the mediation will determine how effectively the negotiations at the mediation proceed. There are three areas of preparation to which close attention should be paid. They are discussed below.

(a) Briefs

The mediation brief is the backbone of the process. It provides the mediator a strong framework for doing his or her preparation, as well as setting out a persuasive case for the other parties. Be as focused and succinct as possible — there is a reason they are called "briefs". Include only what is necessary to cover the key points, and exchange the briefs among all parties. Third party claimants, for example, often lack key information that has been exchanged between the plaintiff and defendant. They should be brought fully up to speed to be an effective party in the mediation. Key parts of the brief include:

- *Fact Summary*: Here, the KISS principle applies — Keep It Simple, Stupid. Clearly cover the relevant and related facts, and support the facts with only the documents necessary to support the facts. If the brief is accompanied with every piece of paperwork generated by the project in three boxes, it will not help anyone move the process forward. On the other hand, if briefs do not contain any supporting documents relevant to the claim, the other parties will pay little attention to the claim itself.
- *Chronology of Key Events*: Create a clear timeline with relevant events filled in. This will help everyone get clear on "exactly what happened" collaboratively, and not waste valuable time at the mediation trying to sort out the basic sequence of events.
- *Applicable Law*: Cover the applicable law, with an explanation of how the legal arguments relate to the law. Support the legal argument only where it is needed. Attaching every construction law case summary of the last 10 years will only waste time. Citing one or two relevant cases when unique arguments are being made can be a major help.
- *Avoid Positions, Focus on the Goals and Objectives of the Client*: This is most often missing. Outline clearly what your party hopes to achieve and why, staying as non-positional as possible. Positions and demands simply provoke defensiveness, whereas goals and objectives promote discussion. Saying "My client demands $1.75 million dollars in damages", will be easily dismissed; saying, "My client needs to be reimbursed for costs that he did not cause", opens the discussion up.
- *Construction Documents*: Be selective. Attach and highlight what is relevant only. Relevant documents can include any contracts, invoices, even blue prints. Be careful, however, not to turn the mediation into a discovery or a hearing. Many hours can be spent pouring over blue prints to little avail. Attach full documents when needed, but excerpt them in the brief to show the relevance to the claim or the issues in dispute.

(b) Client Preparation

Clients are well versed in the strengths of their case; they must come prepared to see the weaknesses of their case, as well as the risks and consequences of not settling. Typically, clients think about and reinforce their own view of the facts, often omitting what they contributed to the problem. In addition, counsel frequently reinforces the strong points of the case with their client, especially early in the claim. Coming to mediation requires that the other side of the coin be examined, and the client be prepared to look hard at the risks and challenges with going forward. In addition, the important non-legal issues should be discussed, such as future relationships in the industry, future projects, negative publicity, *etc*.

Counsel should also plan to have the client directly involved with the mediation. Mediation may be the best (and sometimes only) "day in court" for the client in terms of being able to tell his or her story, and regardless of whether the matter settles, this gives the client an opportunity to be an integral part of the process. The more involvement the clients have, the more likely, in general, that the matter will settle.

Counsel and clients should agree ahead of time when the client will speak, what the client will address, and what he or she should refer to the lawyer to handle. By establishing this during the preparation, both parties are more comfortable and effective when face-to-face with the other parties.

(c) Who Attends

This is critical to the success of the process, as mentioned in Chapter 5. Some specific best practices around this include:

- *Authority*: Since the clients are attending, the most senior person with the most authority to negotiate should attend. When insurance policies are triggered and insurance representatives are attending, counsel should do whatever they can to ensure that the insurance representative brings significant authority to the table, not just nominal or nuisance levels. This will prevent feelings of "bad faith" later in the day when real numbers are being discussed, and will avoid the need for frantic calls being made to try and find someone with authority to close the deal.
- *Representation*: If there are insurance coverage issues, an issue of representation arises. While in most cases the insurance company retains counsel to represent the client, it often reserves the right to try and void the policy at a later date, demanding the full amount of what it paid out from its own client. If this is the case, a party may need to retain its own counsel to advise on the possibility that the insurance company will go after him or her for the loss after settling with the other side.
- *Expert Advice*: In most cases, experts will have submitted written reports and will not attend personally. This can result in each party simply relying on its own expert report, and rejecting the other party's expert opinion, leading to a stalemate. Where the input from experts is

critical, consider arranging for one, both or joint experts to attend to help the parties assess the expert data and assist in their risk assessment and decision making.
- *Direct Knowledge*: Where issues of fact are at play, have the project manager who was on site attend. If financial issues are important, have the accountant or controller attend. Nothing wastes time more than having senior lawyers and senior management sitting in a room arguing what actually happened when the people who were there are not in attendance.

Ask the mediator to manage this issue and ensure that the appropriate parties from all sides are present.

III. HABIT #3: SEE THIS AS A BUSINESS DECISION, NOT A LEGAL PROBLEM

Once a problem enters the legal system, the client has less and less of a role to play in the process. For mediation, this needs to be reversed. Everyone should remember that the clients are the decision makers, the clients must live with the outcome, and therefore the entire process should be focused around helping the clients make a good decision on the matter. In other words, the problems should be treated as less of a legal dispute and more as a business dispute. And while quality legal advice is critical to helping the clients make a good business decision, it should be remembered that a good business decision is the goal, and not the other way around. By doing this, clients will be more involved and will have full ownership of the resolution. Clients who feel they have been an integral part of the process will be empowered to be flexible when necessary to reach agreement, and will take ownership of the outcome if they do not reach agreement. Both outcomes are "wins" for the client, as well as the lawyer.

IV. HABIT #4: USE VISUAL AIDS

Mediation is a flexible process designed to meet the interests of the parties, not the rigid requirements of due process. For that reason, many creative approaches can be incorporated to help the mediation process, including:

- *Photographs*: Take pictures of key areas on the jobsite when it will help everyone understand the problems under discussion. Just remember to include easy-to-understand labels, reference points, scale measurements, *etc.*, so the photos add useful data to the process.

Pictures shot up close with no scale reference or pictures that cause argument about what jobsite they were taken on only waste time.
- *Video*: Video, like photographs, can either add clarity and reality to the parties' understanding of the problem, or it can simply waste everyone's time. Include clear reference points and location markers for starters. Only use video if it demonstrates something clear and directly related to some part of the claim. General walkthroughs of the jobsite are about as interesting as vacation home movies.
- *Charts and Graphs*: Well-designed charts and graphs can summarize complex and voluminous data very effectively. Keep them in proportion to the claim, however. If the mediation is scheduled for one day, a three-hour PowerPoint presentation in the opening statement will not be welcome.

V. HABIT #5: USE THE MEDIATOR EFFECTIVELY

To begin with, different mediators offer very different styles and approaches when it comes to evaluating the case and offering their opinion to the parties. If parties are primarily looking for an evaluation of the case, they should consider a process other than mediation, such as non-binding arbitration or expert neutral opinion. That said, a good mediator is there to facilitate the parties' negotiations, and some well-timed input on how strong certain parts of the case are for either side can be a help. Any input by the mediator about the strength or weakness of either party's case should be a last resort, however, because it can impair the mediator's effectiveness. If a party suddenly feels the mediator does not like his or her case, the party may question the neutrality of the mediator from then on. Once a mediator has lost credibility or impartiality in any parties' eyes, the mediator will be little help in resolving the claim.

What a mediator can do effectively is coach the parties around the effect a particular offer may have, and how it will affect the negotiation process. Remember, the mediator is the only person who is speaking to all the parties throughout the process. The mediator occupies a privileged position, and even when he or she cannot reveal exactly what is going on in each room, he or she can guide the parties toward the most effective direction everyone can take to reach resolution. Parties should look for mediators who are effective at managing the process, coaching the parties, and keeping all parties participating, talking, and moving, as this is the largest predictor of success in the mediation process. In construction claims, mediators who are also familiar and comfortable with the construction field will be better at this than mediators who are not.

VI. HABIT #6: THINK OUTSIDE THE BOX

The bad news about construction disputes is that they involve numerous parties and tend to be complex and detailed. The good news about this complexity is that it opens the door to creative, non-traditional resolutions. In many lawsuits, the only question the parties can focus on is the simple one of "who pays how much money to who?" And while many of these cases settle, they are distributive in nature and revolve around bargaining until a number is reached.

In construction disputes, the parties and the mediator should look for business solutions that go beyond simply writing a cheque. Some of the possibilities in construction disputes include:

- *Commitment to early completion*: One party could invest additional resources in early delivery of the project in lieu of cash payments;
- *Upgrades/Additional Work*: Parties could agree to provide additional value at no cost or a reduced cost as part of the resolution. This can be high value for one party, and lower cost for the other;
- *Future Business*: Parties could agree on being retained for an upcoming contract as settlement of claims on this project. Alternatively, a party could be placed on a shortlist for another contract as part of the settlement.
- *Innovative Work-arounds*: Once a problem has emerged, the tendency is to claim the cost of starting over and redoing the work as per the spec. Another approach is to design a solution that accepts the project as currently built, and "works around" the problem in a way that still delivers the outcome the parties want.

In one case, for example, concrete on a bridge was poured with the wrong additive to compensate for temperature, and resulted in shrinkage that allowed water to pour through the surface. This was unacceptable to the client, even though it did not impact the structural integrity. The claim was for tearing the concrete out completely and redoing the work, at huge expense. In mediation, the agreement was to try a new product that filled and sealed the gaps, and extending the warranty period to compensate for uncertainty that comes with using a new product.

Creative solutions require equally creative ways of documenting the final settlement to ensure future disputes are not created along the way, as noted below.

VII. HABIT #7: DOCUMENT THE SETTLEMENT WELL

Resolving complex, high risk disputes is hard on the parties, and once an agreement is reached in principle there is a strong tendency for parties to either want to shake hands and leave before the deal falls apart, or to write up the main points and leave the rest for counsel to worry about. This is a mistake.

Parties need to take the time to document or memorialize their agreement in such a way that it becomes a legally binding contract. If the settlement is simply monetary, it is pretty straightforward. If the settlement is more creative and complex, such as requiring specific performance of certain things within certain timeframes, parties need to "operationalize" the agreement, walking through each step to make sure it meets everyone's expectations. In cases where the project is still ongoing, agreements on how issues will be prevented in the future must be discussed. If this is not done, it simply invites more conflict later on.

In all cases, thought should be given to what will happen if any party does not perform on the agreement. In some cases, parties can rely on the bonding company to step in, but that often opens another area of dispute. Another way is to agree on stipulated damages or consent judgments in the event of default, thus increasing the likelihood parties will perform.

Regardless of what direction the parties go in reaching the agreement, all parties must bring the same commitment to documenting the deal as they did to reaching the deal. To paraphrase a great thinker "The mediation ain't over 'til it's over", and it is not over until the documentation is done well.

VIII. SUMMARY

These are some of the specific best practices to consider when mediating construction disputes. Because of the complexity and magnitude of many construction claims, parties will be well served to address these issues well in advance of the mediation. The more experienced counsel becomes with mediating construction claims, the more likely it is that these best practices will be used to set up and conduct a mediation that has the best possible chance of reaching a resolution.

7
THE FUTURE OF CONSTRUCTION ADR

In considering some of the best practices around construction dispute resolution, three principles should be clear:

- First, parties need to consider their dispute resolution strategy and approach from the beginning of the project, for the simple reason that conflict and disputes are a virtual certainty on a construction project. The more effectively the parties think about and plan for how disputes will be resolved, the less chance there is that conflict will cause the project problems and delays.
- Second, once a dispute has arisen, parties need to continually frame the problem as a business issue rather than as a legal issue. In other words, the legal advice should be in service of the clients making a good business decision, rather than the legal issues controlling and driving the decision through the courts. The more effectively parties can stay focused on what they really want (and few clients really want a lawsuit!) and on their interests, the more likely it is that a solution that keeps the project moving and the construction moving ahead will be found. Part of this principle includes addressing issues immediately, as soon as they arise. Avoiding the problem, sitting back and "papering" the file to strengthen the legal case, or taking hard line positions and refusing to talk all have the effect of delaying resolution. This, of course, simply increases the delays and the damages, and makes the entire problem harder to resolve.
- Third, parties should sequence their dispute resolution approach to start with interest-based processes like negotiation and mediation first, as the first step in the conflict resolution process. Only if mediation fails should the parties move to rights-based processes such as arbitration or litigation. Remember, 97 per cent of lawsuits settle through some type of negotiation, so the sooner parties negotiate or mediate the issues, the more effectively they will resolve the conflict and move on.

By focusing on these key principles and applying some of the best practices put forward in this book, disputes will resolve faster and cheaper while maintaining the parties' working relationship as much as is possible.

I. THE FUTURE OF ADR IN CONSTRUCTION DISPUTES

The future direction of dispute resolution in the construction industry is hard to identify definitively, as there are no studies or sources of data that point in any conclusive direction. That said, there are a number of trends that anecdotally point strongly to the increased use of the full range of ADR processes. Effective dispute resolution in the construction industry is gaining ground and momentum, for the simple reason that the traditional litigation route takes too long and costs too much, both in terms of money and in terms of relationships. The construction industry is looking for better ways to find out what went wrong, find out why, find out the cost of fixing it, actually fixing it and moving on in the fastest, least expensive way possible. ADR is being chosen more often to help deliver this. Below are some of the trends that will affect the use of ADR in the future.

(a) Government Intervention

ADR is being adopted by a number of jurisdictions in a variety of ways all across North America. Led by the state of Florida, many U.S. states have a form of mandatory mediation for a wide range of litigation disputes, and construction disputes are frequently included in these schemes. There is no question that the courts have recognized the ability of mediation to both reduce the amount of litigation before the courts and to give the parties faster and cheaper ways of settling lawsuits.

In Ontario, the provincial government has committed itself to implementing the Ontario Mandatory Mediation Program throughout the province. As of today, three jurisdictions (Toronto, Ottawa and Windsor) require mediation early[1] in most lawsuits.[2] In addition, the government is on record that it wishes to include construction lien matters in mandatory mediation, but it has currently taken few steps to make that a reality. In the automobile insurance area of the law, whenever Accident Benefits disputes arise parties are required to mediate (and can later choose to arbitrate) through the Financial Services Commission of Ontario, an

---

[1] At the moment, mediation is required to take place within 90 – 150 days of the filing of the first defence, typically prior to production or discoveries. Parties do have the ability to request, on consent, extensions to allow discovery in any given file prior to mediation, but a case management master must grant this request.

[2] Currently, all lawsuits, with the exception of family, commercial list, and construction lien matters are mandated to mediation. Simplified Rules cases are mandated into mediation in Ottawa, and excluded from mandatory mediation in Toronto.

agency that employs over 40 mediators and 10 arbitrators on a full time basis.

In British Columbia, the approach they have taken does not require that lawsuits be mediated in all cases. Under its rules, if either party requests mediation in a case, the other party must comply and participate. In other words, British Columbia has a voluntary approach to mediation, which becomes mandatory if either party wishes mediation.

In Saskatchewan, the government employs a small number of mediators who take cases referred selectively from the bench. In Alberta, judges have taken a leading role in providing some mediation and mini-trial services to larger files, on a case-by-case basis.

Specifically in construction lien cases, Master David Sandler of the Superior Court of Ontario has been working almost exclusively with the case management of construction lien matters for many years. Master Sandler identified three types of cases that come before him as Master of construction lien matters:

- True construction lien disputes, focused on issues such as the procedural validity of a lien, timeliness, priority of claims under a lien action, quantum, *etc.*
- Construction law disputes, focused on issues such as fiduciary duty, unjust enrichment, *etc.*
- Factual disputes in a construction matter, focused on issues such as quality, scope of contract, deficiencies, specifications, *etc.*

According to Master Sandler, while it is the factual issues category that is most amenable to the use of mediation, all three would benefit from the use of both ADR and early fixed trial dates. In relation to early trial dates, Sandler has found that parties tend to focus much earlier on settlement and resolution the sooner a fixed, unmovable trial date is established. In relation to the idea of mandatory mediation, Sandler stated that he would "certainly have mandatory mediation in all lien cases", adding that mediation should come after production of documents in all cases, and after discovery or at least limited discovery in many of the cases. Mediation works, according to Sandler, in part because "the right mediator has credibility with the parties more often than the courts do". In addition, Master Sandler added that mediation could address a much wider range of issues. "In the courts, there are limits to what we can do", said Sandler. "In ADR, there aren't really any limits to what you can to do reach a resolution".

Master Sandler's views are consistent with the views of many of the people who run the civil justice system. It appears that the future direction of dispute resolution in the construction industry is the growing use of various forms of ADR, with a strong emphasis on mediation. In the Civil

Justice Review of 1996 conducted by the Hon. Robert Blair, the report looked at construction lien cases, and recommended: "construction lien cases, like other civil cases, should be subject to mandatory referral to mediation after a defence has been filed".[3] Based both on these strong "nudges" from the legal system, as well as from positive results from many parties who have used ADR, many people in the construction industry are becoming knowledgeable and skilled in the use of ADR.

(b) Partnering and Other Preventative Approaches

(i) Partnering

One of the strongest indicators of the movement toward proactive use of ADR has been the adoption of Partnering in Canada on large construction projects. Partnering is a relatively new process, having been developed in the United States under 20 years ago. Since its inception and use by the Army Corps of Engineers, Partnering has been embraced and used around the world in a very short time. In Canada, Partnering was introduced just over 10 years ago, and has been implemented successfully and frequently by large project owners and contractors across the country, including the Department of National Defence, Public Works and Government Services, the Ministry of Transportation in Ontario, PCL Constructors, Ellis-Don, and many municipalities and private builders. In fact, PCL found the process so helpful that it trained its own Partnering facilitators and began Partnering all projects with a budget of approximately $10 million or over in size.[4]

While there appears to be a small decline in the use of Partnering in the last few years, the reason for this appears to be that on some projects, the participants have been through Partnering so many times that they bring the knowledge, skills and attitudes necessary for good collaboration and dispute resolution to the table without the need for a workshop. If anything, this supports the value and success of interest-based approaches to managing conflict.

---

[3] *Ontario Civil Justice Review*, First Report (March 1995), Supplemental and Final Report (November 1996), The Hon. Mr. Justice Robert A. Blair and Susan Lang, Assistant Deputy Attorney General, Co-Chairs.

[4] While in most Partnering processes the facilitators are "neutral" facilitators, *i.e.*, not part of any company related to the project, PCL has had good success using internal facilitators to run its Partnering sessions.

(ii) Public/Private Partnerships

The industry is already seeing new and innovative ADR processes taking shape. One of the strongest movements in many governments today is the idea of Public/Private Partnerships, or P-3s. These are projects where the public sector creates a partnership with a private company or consortium to build and manage a facility for a significant period of time, frequently in the 25-year (or longer) range. On a recent call for proposals to build and manage a major health care facility structured as a P-3, the private bidder included a full dispute resolution system, along with a dispute resolution provider, as an important part of their bid. The dispute resolution professional was mandated to design and facilitate good dispute resolution not only during the building of the facility, but also for the full term of the agreement. Clearly, the construction industry itself is beginning to see the ongoing value of good long-term dispute resolution.

(iii) Training and Skills

In the past, the primary skill set for a good project manager was knowledge of construction processes, and perhaps the ability to be tough enough to get the job done. This is changing. In the past few years, good conflict resolution skills are being added to the list for construction project management positions. One major contractor has undertaken to train all of its project managers across the country in effective conflict management skills. In addition, the Project Manager's Institute has included conflict resolution training as part of its continuing education offerings. More and more, effective ADR and the related skill set are hitting the radar screen of the construction industry.

II. SUMMARY

If everything always went according to plan, there would be no need for ADR. The reality, though, is that the only constant is change, and the construction industry deals with more change than most — change to the plans, change in the weather, change in the expected soil conditions, change in personnel, change in schedules, and many, many more. And the fact, quite simply, is this: Change causes conflict. The people and organizations that can better manage and resolve conflict will accomplish change more effectively, more quickly and more cheaply.

The more effectively everyone in the construction field applies ADR Best Practices before, during and after each and every project that is built, the better the outcomes will be for everyone concerned.

Appendix 1

BIBLIOGRAPHY

Armstrong, David T., "Maintaining Control of a Construction Project Even When Problems Arise" *Just Resolutions* (October 2000).

Banfai, Geza R., "Why Construction Lawyers Must Change" (1998) 36 C.L.R. (2d)

Bristow, David I., Q.C. and Seth, Reva, "Good Faith in Negotiations", American Arbitration Association, *Dispute Resolution Journal* (November 2000).

Bristow, David I., Q.C. "The New CCDC2: Facilitating Dispute Resolution of Construction Projects", Canadian Bar Association, Ontario lecture, (December 1998).

Chaykin, Arthur A., "Selecting The Right Mediator", American Arbitration Association, *Dispute Resolution Journal* (1994).

Crowley, Thomas E., *Settle It Out of Court: How to Resolve Business and Personal Disputes using Mediation, Arbitration and Negotiation* (John Wiley & Sons Inc., 1994).

Curtis, Gordon H., "Improving the Outcome of Mediation: 15 Tips You Can Count On", American Arbitration Association, *The Punch List,* Vol. 22, No. 1 (May 1999).

Fenn, Peter, "Mediating Building Construction Disputes", *Rethinking Disputes : The Mediation Alternative,* Julie Macfarlane, ed. (Emond Montgomery Publications Limited, 1997).

Fisher, Timothy S., "Construction Mediation", American Arbitration Association, *Dispute Resolution Journal* (March 1994).

Glaholt, Duncan W. and Rotterdam, Marcus, "Toward an Inquisitorial Model for the Resolution of Complex Construction Disputes" (1998) 36 C.L.R. (2d) 129.

Glaholt, Duncan W., *Conduct of a Lien Action* (Toronto: Carswell, 2004).

Golann, Dwight, "Mediating Legal Disputes: Effective Strategies for Lawyers and Mediators", *Aspen Law and Business* (2001), at 36.

Hart, Ross R., "Improving your Chance of Success during Construction Mediation", American Arbitration Association, *Arbitration Journal* (December 1992).

Hinchey, John W., "A Wake-Up Call to the Construction Bar" (1997) 2 Construction Forum News, No. 2.

Kiel, James H., "Hybrid ADR in the Construction Industry", American Arbitration Association, *Dispute Resolution Journal* (August 1999).

Marston, Donald L., "Project-Based Dispute Resolution: ADR Momentum Increases into the Millennium" (1999) 48 C L.R. (2d) 221.

Martell, Ronald E. and Patterson, John G., "Practical Aspects of Mediating Construction Disputes", American Arbitration Association, *The Punch List* (January 1994).

McGovern, Francis E., "Strategic Mediation: The Nuances of ADR in Complex Cases", American Arbitration Association, *Dispute Resolution Magazine* (Summer 1999).

McGuire, James E., "Practical Tips for Using Risk Analysis in Mediation", American Arbitration Association, *Dispute Resolution Journal* (May 1998).

Medved, George M., "Using ADR to Resolve Delay Claims Cases", American Arbitration Association, *The Punch List,* Vol. 15, No. 3 (September 1992).

Moore, Christopher *The Mediation Process*, 3rd ed. (San Francisco: Jossey-Bass Publishers, 1986).

Myers, James J. "Resolving Disputes in Worldwide Infrastructure Projects" (1999) 47 Construction L. Rev. (2d) 87.

Myers, James J., "Survival Kit for Complex Construction Arbitration" American Arbitration Association, *Dispute Resolution Journal* (September 1994).

Naughton, Philip, Q.C., "Alternative Forms of Dispute Resolution — Their Strengths and Weaknesses" (1990) 56 J.C.J. Arb. 76, *Construction Law Journal* (U.K.).

Ossman III, George ("Chip"), "Construction Arbitration: A Survey to Arbitrator's Award Consistency" (2002) Construction Management Association of America CM eJournal, <http://cmaanet.org/user_images/arbitration_paper_ossman.pdf>.

Schrader, Charles R., "Construction Mediation: Why and How", American Arbitration Association, *The Punch List* (Winter 1994).

Smith, John A., "Construction ADR: You Get Out What You Put In", American Arbitration Association, *Dispute Resolution Journal* (July/September 1995).

Venzie Jr., Howard D., "Closure Issues in Construction Mediation", American Arbitration Association, *The Punch List* (November 2000).

Wilkinson, John, "Streamlining Arbitration of the Complex Case" (August/October 2000), American Arbitration Association, *Dispute Resolution Journal*.

Zaghloul, Ramy and Hartman, Francis, "Construction Contracts and Risk Allocation", Proceedings of the Project Management Institute Annual Seminar & Symposium (October 2002).

Advisory Committee on the Alternative Resolution of Construction Disputes, "Too Many Disputes! Too Much Litigation! Dispute Resolution Opportunities for the Construction Industry", Province of Ontario, Ministry of the Attorney General (June 1994).

PARTNERING BIBLIOGRAPHY

Carr, Frank, "Partnering: Dispute Avoidance, The Army Corps of Engineers Way", American Arbitration Association, *The Punch List*, Vol. 14, No. 3 (1991).

Cook, Lynn E. and Hancher, Donn E., "Partnering: Contracting for the Future" (1990), Vol. 6, No. 4 J. of Management in Engineering.

Pinnell, Steve, "Partnering and the Management of Construction Disputes", American Arbitration Association, *Dispute Resolution Journal* (February 1999).

Warne, T.R., "Partnering for Success" (New York: ASCE Press, 1994).

Appendix 2

DRAFT AGREEMENT TO MEDIATE

Short Style of Cause:

File #:

Date of Mediation:

The signing of this document is evidence of the agreement of the parties, their counsel and the mediator to conduct this mediation process in a *bona fide* and forthright manner and to make a serious attempt to resolve the outstanding matters.

1. The mediator is a neutral facilitator who will assist the parties to reach their own settlement. The mediator does not offer legal advice. The mediator has no duty to assert or protect the legal rights of any party, to raise any issue not raised by the parties themselves or to determine who should participate in the mediation. The mediator has no duty to ensure the enforceability or validity of any agreement reached.
2. The parties attending the mediation will have full, unqualified authority to reach a settlement in this matter.
3. It is understood that in order for mediation to work, open and honest communications are essential. Accordingly, all written and oral communications, negotiations and statements made in the course of mediation will be treated as without prejudice settlement discussions and are absolutely confidential. Therefore:
    (a) The mediator will *not* reveal anything discussed in mediation without the permission of all parties.
    (b) The parties agree that they will not at any time, before, during, or after mediation, call the mediator as witnesses in any legal or administrative proceeding concerning this dispute. To the extent that they may have a right to call the mediator as witnesses, that right is hereby waived.
    (c) The exception to the above is that this agreement to mediate and any written agreement made and signed by the parties as a result of mediation may be used in any relevant proceeding, unless the parties make a written agreement not to do so.
    (d) The mediator may discuss the mediation in a manner that does not in any way reveal the identity of the parties or the case for educational and teaching purposes only.
4. It is understood that disclosure of all relevant and pertinent information is essential to the mediation process. Accordingly, there will be complete and honest disclosure by each of the parties to the other and

to the mediator of all *relevant* information and documents. This includes providing each other and the mediator with all information and documentation that usually would be available through the discovery process in a legal proceeding.
5. While all parties intend to continue with mediation until a settlement agreement is reached, it is understood that any party may withdraw from mediation at any time. It is agreed that if any party decides to withdraw from mediation, best efforts will be made to discuss this decision in the presence of all parties and the mediator.
6. If the mediator determines that it is not possible to resolve the issues through mediation, the process can be terminated once this has been conveyed to the parties.
7. It is agreed that where a settlement is reached in the dispute, the parties and their counsel will carry out the terms of the settlement as soon as possible.
8. It is agreed that the mediator's fee will be charged at the rate of $ XXX.00 per hour for any portion of an hour, with a minimum billing of four (4) hours, split between the parties, plus applicable taxes and applicable disbursements. Accounts unpaid after 30 days will be charged interest at two (2) per cent per month.
9. Counsel hereby agree and undertake to pay the mediator's fees, which fees shall be divided equally between or among the parties unless otherwise agreed.
10. It is agreed that the parties are responsible for fees and expenses associated with any adjournment or cancellation of the mediation once the date is booked. The cancellation fee is equal to the minimum billing rate of four hours plus any disbursements incurred, if cancellation occurs within seven working days of the date established.
11. This Agreement may be executed by the parties in separate counterparts each of which when so executed and delivered shall be an original, but all such counterparts shall together constitute one and the same instrument. This Agreement shall ensure to the benefit of and be binding upon the parties hereto, their heirs, executors, administrators, successors and assigns.

Dated at _____ this _____ day of _____, 200_____

_____   _____

Plaintiff by his/her Counsel   Defendant by his/her Counsel

_____   _____

Plaintiff by his/her Counsel   Defendant by his/her Counsel

_____   _____
Mediator   Mediator

Appendix 3

TOWN OF NEW BEDFORD / ONTARIO WATER SUPPORT AGENCY AND OAK ENGINEERING SEWAGE INFRASTRUCTURE UPGRADES OCTOBER 29 & 30, 2001

PARTNERING CHARTER

Motto — "Partners in Excellence"

(Note – Team Photo Here)

Everything we do will be consistent with our commitment to do the best job possible for the benefit of the people of New Bedford and the Partnership. To this end, we commit to:

1. Work together to create a facility that is operator-friendly and offers low operation and maintenance costs to the municipality;
2. Complete the project on time and on or under budget;
3. Demonstrate innovation and apply the expertise of the team to meet its objectives;
4. Create and maintain a positive image for the project with the Town of New Bedford and with the public generally;
5. Support decision-making responsibility of representatives in the field to move the project forward;
6. Work as a team to maximize the value of any changes to the ultimate success of the project;
7. Construct a state of the art facility using the best available technology economically achievable;
8. Create a team with the expertise to take on similar projects elsewhere;

9. Produce an excellent project that will become a benchmark for projects of its kind in the future;
10. Maintain the partnering approach and avoid conflict;
11. Produce a quality project that delivers value to all members of the partnership;
12. Do the job well and have fun doing it.

Appendix 4

CONTRACTUAL DISPUTE RESOLUTION CLAUSES[*]

The dispute resolution clauses in the standard contracts from CCDC and CCA are all handled in a similar fashion. What is also common to all of them is the strong recommendation from these organizations that parties use the dispute resolution clauses to minimize the costs of dealing with disputes. The following excerpt is from the introduction to the CCDC-40 – the Rules for Mediation and Rules for Arbitration of CCDC-2 Construction Disputes:

> *Construction disputes are common. The multiplicity of parties and technical complexity of major construction projects makes these projects very susceptible to disagreements. The costs associated with resolving major disputes is burdensome even for the largest organizations. Delay in resolving outstanding disputes causes serious cash flow problems for smaller companies and sub-trades. Acrimony fed by the adversarial system can seriously impair or permanently damage future business dealings. When you add this to the problems of overburdened courts, it is easy to understand the dissatisfaction of Owners, Contractors, and Consultants with lawyers, judges and the courts and it is easy to understand why players in the construction industry are seriously searching for more effective ways to manage conflict on construction projects.*
>
> *CCDC encourages national use of these Rules for Mediation and Arbitration in the construction industry."*[1]

In the CCDC 2 (stipulated price contract), the timing and major steps in the dispute resolution process are included in the CCDC 2 contract document itself, with reference to the CCDC 40 — Rules for Mediation and Arbitration of CCDC 2 Construction Disputes, which contains the detailed rules themselves for how the steps will be implemented. In addition, the CCA 1 (stipulated price subcontract) incorporates the CCDC 40 by simply changing the word "contract" to the word "subcontract" throughout the CCDC 40. Since the CCDC 40 is a detailed document which bears reading in full, we have reproduced

---

[*] Reproduced with permission of Canadian Constuction Documents Committee.
[1] These Mediation and Arbitration Rules for CCDC 2 Construction Disputes are published and distributed by CCDC under license from Bonita J. Thompson, Q.C., copyright 1993. These Rules may not be reproduced in whole or in part without the written permission of CCDC.

with permission the dispute resolution clause from the CCDC 2 contract to give you a sense of how the contract clause is written. Full copies of these contracts can be obtained directly from CCDC.[2]

---

[2] Canadian Construction Documents Committee, 400 – 75 Albert Street, Ottawa Ontario K1P 5E7.

Part 8 DISPUTE RESOLUTION

GC 1 8.1 AUTHORITY OF THE CONSULTANT

8.1.1 Differences between the parties to the *Contract* as to the interpretation, application or administration of the *Contract* or any failure to agree where the agreement between the parties is called for, herein collectively called disputes, which are not resolved in the first instance by findings of the *Consultant* as provided in GC 2.2 – ROLE OF THE CONSULTANT, shall be settled in accordance with the requirements of Part 8 of the General Condition – DISPUTE RESOLUTION.

8.1.2 If a dispute arises under the *Contract* in respect of a matter in which the *Consultant* has no authority under the contract to make a finding, the procedures set out in paragraph 8.1.3 and paragraphs 8.2.3 to 8.2.8 of GC 8.2 – NEGOTIATION, MEDIATION AND ARBITRATION, and in GC 8.3 – RETENTION OF RIGHTS apply to that dispute with the necessary changes in detail as may be required.

8.1.3 If a dispute is not resolved promptly, the *Consultant* shall give such instructions as in the *Consultant's* opinion are necessary for the proper performance of the *Work* and to prevent delays pending settlement of the dispute. The parties shall act immediately according to such instructions, it being understood that by so doing neither party will jeopardize any claim the party may have. If it is subsequently determined that such instructions were in error or at variance with the *Contract Documents*, the *Owner* shall pay the *Contractor* costs incurred by the *Contractor* in carrying out such instructions which the *Contractor* was required to do beyond what the *Contract Documents* correctly understood and interpreted would have required, including costs resulting from interruption of the *Work*.

GC 8.2 NEGOTIATION, MEDIATION AND ARBITRATION

8.2.1 In accordance with the latest edition of the Rules for Mediation of CCDC 2 Construction Disputes[3], the parties shall appoint a Project Mediator
    .1 within 30 days after the *Contract* was awarded, or

---

[3] The Rules for Mediation and Arbitration of CCDC 2 Construction Disputes are collectively called the CCDC 40.

.2 if the parties neglected to make an appointment within the 30 day period, within 15 days after either party by notice in writing requests that the Project Mediator be appointed.

8.2.2 A party shall be conclusively deemed to have accepted a finding of the *Consultant* under GC 2.2 – ROLE OF THE CONSULTANT and to have expressly waived and released the other party from any claims in respect of the particular matter dealt with in that finding unless, within 15 *Working Days* after receopt of that finding, the party sends a notice in writing of dispute to the other party and to the *Consultant*, which contains the particulars of the matter in dispute and the relevant provisions of the *Contract Documents*. The responding party shall send a notice in writing of reply to the dispute within 10 *Working Days* after receipt of the notice of dispute setting out the particulars of this response and any relevant provisions of the *Contract Documents*.

8.2.3 The parties shall make all reasonable efforts to resolve their dispute by amicable negotiations and agree to provide, without prejudice, frank, candid and timely disclosure of relevant facts, information and documents to facilitate these negotiations.

8.2.4 After a period of 10 *Working Days* following receipt of a responding party's notice in writing of reply under paragraph 8.2.2, the parties shall request the Project Mediator to assist the parties to reach agreement on any unresolved dispute. The mediated negotiations shall be conducted in accordance with the latest edition of the Rules for Mediation of CCDC 2 Construction Disputes.

8.2.5 If the dispute has not been resolved within 10 *Working Days* after the Project Mediator was requested under paragraph 8.2.4 or within such further period agreed by the parties, the Project Mediator shall terminate the mediated negotiations by giving notice in writing to both parties.

8.2.6 By giving notice in writing to the other party, not later than 10 *Working Days* after the date of termination of the mediated negotiations under paragraph 8.2.5, either party may refer the dispute to be finally resolved by arbitration under the latest edition of the Rules for Arbitration of CCDC 2 Construction Disputes. The arbitration shall be conducted in the jurisdiction of the *Place of the Work*.

8.2.7 On expiration of the 10 *Working Days*, the arbitration agreement under paragraph 8.2.6 is not binding on the parties and, if a notice is not given under paragraph 8.2.6 within the required time, the parties may refer the unresolved dispute to the courts or to any other form of dispute resolution, including arbitration, which they have agreed to use.

8.2.8 If neither party requires by notice in writing given within 10 *Working Days* of the date of notice requesting arbitration in paragraph 8.2.6 that a dispute be arbitrated immediately, all disputes referred to arbitration as provided in paragraph 8.2.6 shall be

.1 held in abeyance until
  (1) *Substantial Performance of the Work*,
  (2) the *Contract* has been terminated, or
  (3) the *Contractor* has abandoned the *Work*,
  whichever is earlier, and

.2 consolidated into a single arbitration under the rules governing the arbitration under paragraph 8.2.6.

GC 8.3 RETENTION OF RIGHTS

8.3.1 It is agreed that no acto by either party shall be construed as a renunciation of waiver of any rights or recourses, provided the party has given the notices required under Part 8 of the General Conditions – DISPUTE RESOLUTION and has carried out the instructions as provided in paragraph 8.1.3.

8.3.2 Nothing in Part 8 of the General Conditions – DISPUTE RESOLUTION shall be construed in any way to limit a party from asserting any statutory right to a lien under applicable lien legislation of the jurisdiction of the *Place of the Work* and the assertion of such right by initiating judicial proceedings is not to be construed as a waiver of any right that the party may have under paragraph 8.2.6 to proceed by way of arbitration to adjudicate the merits of the claim upon which such a lien is based.

INDEX

A

Address of project, 36
Alternative dispute resolution
- arbitration. *See* **Arbitration**
- conflict escalation, 18-21
- conflict poison, 19-21
  - fall back to positional bargaining, 21, 26
  - hidden agendas, 21, 26
  - reactive devaluation, 20, 26
  - reversal of
    - barriers re, 19
    - interest-based processes re, 24
  - substantive vs. relationship issues, 20, 25
- conflict prevention. *See* **Conflict prevention**
- dispute resolution stairway, 21-24
  - binding processes, 24
  - litigation, 24
  - negotiation, 23
  - non-binding processes, 23
  - prevention, 22
  - standing neutral, 23
- future developments. *See* **Future of construction ADR**
- importance of, 2
- interest-based processes, 15-16, 18-21
- litigation. *See* **Litigation**
- mediation. *See* **Mediation**
- neutral evaluation. *See* **Neutral evaluation**
- power-based processes, 17-21
- practices after project. *See* **Construction project, ADR practices after**
- practices during project. *See* **Construction project, ADR practices during**
- principled approach to, 105
- rights-based processes, 16, 18-21

Arbitration
- arbitration-mediation hybrid, 88
- arbitrator, choice of, 38
- contract provision re, 38
- described, 81-82
- features of, 81-82
- mediation-arbitration hybrid, 87-88
- non-binding arbitration, 80
- pitfalls of, 82-58
  - appeal right, lack of, 85
  - costs and time, 82-83
  - fairness, 83-85
  - predictability of outcome, 83-85
- practices in, recommended, 85-87
  - evidence, focus on handling of, 86
  - hearing, tight control by arbitrator, 87
  - procedure, initial focus on, 86

B

Bibliography, 111-13
Bonds, bid and performance, 41

C

Change orders, 9, 41, 64-5
Communication
- decision-making vs., 34
- during construction project, 49-59
  - daily site meetings, 50-52
    - sample minutes, 51-52
    - summary re, 51
  - senior management meetings, 57-58
  - weekly site meetings, 52-56
    - attendance, 52
    - meeting agendas, 53
    - minutes, 53-54, 55-56
    - summary, 55
- generally, 10-11, 58-59
- mode of, contract provision re, 42

Conflict poison, *see also* **Alternative dispute resolution**
- examples of, 4-6
  - delay issues, 5
  - disappointing results issues, 6
  - weather issues, 5-6
- "heart" of construction project, 3-4
- origins of, 3-4

Conflict prevention
- case studies re, 43-47
- communication vs. decision-making, 34
- contract provisions. *See* **Contract provisions**
- expectations, discussion between parties re, 27-28
- hiring precautions, 27-28
- issue-resolution process, 33-34
- partnering, 30-33
  - described, 31-33
  - methodology, as, 33
  - origins of process, 30
  - partnering workshop, 32-33
  - statistics re success of, 31
- quotes, detailed discussions re, 28
- references, checking of, 29-30

Construction industry
- complexity of, 3
- conflict poison in. *See* **Conflict poison**
- dispute in. *See* **Disputes**

- inevitability of conflict, 2-3
- interdependence, 7

Construction Lien Act. See **Construction project elements**

Construction project, ADR practices after
- arbitration. *See* **Arbitration**
- case studies, 91-96
- generally, 71
- litigation. *See* **Litigation**
- mediation. *See* **Mediation**
- neutral evaluation. *See* **Neutral evaluation**

Construction project, ADR practices during case studies re, 66-70
- communication. *See* **Communication**
- issue resolution process, 59-66
  - change orders or extras, issues re, 64-65
  - issue becoming dispute, reason why, 59
  - issue resolution ladder, 60-63
  - standing neutral, appointment of, 64
  - value engineering process, issues re, 65-66

Construction project elements
- blue prints, 10
- closing project, 11
- communication and co-ordination, 10-11
- *Construction Lien Act*
  - holdback, 8, 12
  - substantial performance and completion, 11-12
- dates, commencement and completion, 42
- insurance, 41

Contract provisions
- address of project, 36
- bid and performance bonds, 41
- change orders, 41
- communication mode, 42
- dates, commencement and completion, 42
- deficiencies, 42
- delays or disclaimer clauses, 42
- dispute resolution clauses, 36-38, 121-25
  - authority of consultant, 123
  - CCDC draft contracts, 36, 121-25
  - key provisions of, 37
  - mandatory steps, 37
  - negotiation, mediation and arbitration, 123-25
  - retention of rights, 125
- mediator or arbitrator, choice of, 38
- parties, 35
- payment stream and invoicing, 41
- project insurance, 41
- risk allocation, 40
- standing mediator or neutral, appointment of, 39
- value engineering, 42
- warranties, 42

D

Deficiencies, 42
Delay

- conflict poison, as, 5
- contract provision re, 42

Dispute resolution
- contract clauses. *See* **Contract provisions**
- generally. *See* **Alternative dispute resolution**

Disputes
- case studies re, 91-96
- *Construction Lien Act* holdback, 8
- construction project structure and. *See* **Construction project elements**
- generally, 1-2, 6-7
- interest-based, ADR, 13-14
- tendering process, 8
- types of, 7-9
  - change orders, 9
  - payment schedule, 8
  - substandard quality, 9
  - timing, 7-8

E

Expectations, 27-28

F

Future of construction ADR
- generally, 106, 109
- government intervention, 106-108
- Ontario Mandatory Mediation Program, 106
- Partnering, 108
- Public/Private Partnerships, 109
- training and skills, 109

H

Hiring, 27-28

I

Interest-based processes, 13-16, 18-21, 24
Issue resolution. *See* **Construction project, ADR practices during**

L

Litigation
- avoidance of, 90
- complexity of, 90
- costs of, 89-90

M

Mediation
- arbitration-mediation hybrid, 88
- attendance, 100
  - authority, 100
  - direct knowledge, 101
  - expert advice, 100
  - representation, 100

- business decision, as, 101
- creative solutions, importance of, 103
- draft agreement re, 115-17
- generally, 72
- mediation-arbitration hybrid, 87-88
- mediator
  - choice of, 38
  - effective use of, 102
  - substantive knowledge required of, 76-77
- preparation, 98-101
  - briefs, 98-99
  - client preparation, 99-100
- process described, 72-76
  - agreement, documenting, 76
  - introduction and opening statements, 74
  - issues, identifying, 75
  - minutes of settlement, 76
  - pre-mediation considerations, 73-74
  - problem solving, 75-6
- pros and cons of, 77-9
- settlement, documentation of, 103-104
- timing
  - early, 99
  - late, 98
- visual aids, use of, 101-102

Meetings. *See* **Communication**
Mini-trial, 81

N

Neutral evaluation
- generally, 79
- mini-trial, 81
- neutral expert opinion, 80
- non-binding arbitration, 80

O

Ontario Mandatory Mediation Program, 106

P

Partnering, *see also* **Conflict prevention**
- bibliography, 113
- future of, 108
- sample charter re, 119-20

Payment schedule issues
- contract provision re, 41
- disputes re, 8

Positional bargaining, 21, 26
Power-based processes, 17-21

Q

Quotes, 28

R

Rights-based processes, 16, 18-21
Risk allocation, 40

S

Settlement
- documentation of, 103-104
- minutes of, 76

Standing neutral
- appointment of, 39, 64
- dispute resolution stairway, 23

Substandard quality issues, 9

T

Timing issues, 7-8

V

Value engineering, 42

W

Warranties, 42